LIGHT + DESIGN

Cover Photo Credits:

Top – Bismarck Airport, Bismarck, ND
Architect: Tvenge Associates Engineers – Viteig
Lighting – Litecontrol
Photography: Charles Mayer Photography

Center left - Rion Antiron Bridge in Greece
Architect: Berdj Mikaelian
Lighting design: Roger Narboni, CONCEPTO studio, France
Photo copyright: CONCEPTO

Center right – Minneapolis Convention Center
Lighting design: Schuler Shook

Bottom – The Orange County Performing Arts Center
Lighting Design: Ross De Alessi

Light + Design
A Guide to Designing Quality Lighting
for People and Buildings

Quality of the Visual Environment Committee,
Illuminating Engineering Society of North America

Copyright 2008 by the Illuminating Engineering Society of North America.

Approved by the IES Board of Directors, November 9, 2008, as a Transaction of the Illuminating Engineering Society of North America.

All rights reserved. No part of this publication may be reproduced in any form, in any electronic retrieval system or otherwise, without prior written permission of the IES.

Published by the Illuminating Engineering Society of North America, 120 Wall Street, New York, New York 10005.

IES Standards and Guides are developed through committee consensus and produced by the IES Office in New York. Careful attention is given to style and accuracy. If any errors are noted in this document, please forward them to Rita Harrold, Director of Technology, at the above address for verification and correction. The IES welcomes and urges feedback and comments.

ISBN # 978-0-87995-231-0

Printed in the United States of America.

DISCLAIMER

IES publications are developed through the consensus standards development process approved by the American National Standards Institute. This process brings together volunteers representing varied viewpoints and interests to achieve consensus on lighting recommendations. While the IES administers the process and establishes policies and procedures to promote fairness in the development of consensus, it makes no guaranty or warranty as to the accuracy or completeness of any information published herein.

The IES disclaims liability for any injury to persons or property or other damages of any nature whatsoever, whether special, indirect, consequential or compensatory, directly or indirectly resulting from the publication, use of, or reliance on this document

In issuing and making this document available, the IES is not undertaking to render professional or other services for or on behalf of any person or entity. Nor is the IES undertaking to perform any duty owed by any person or entity to someone else. Anyone using this document should rely on his or her own independent judgment or, as appropriate, seek the advice of a competent professional in determining the exercise of reasonable care in any given circumstances.

The IES has no power, nor does it undertake, to police or enforce compliance with the contents of this document. Nor does the IES list, certify, test or inspect products, designs, or installations for compliance with this document. Any certification or statement of compliance with the requirements of this document shall not be attributable to the IES and is solely the responsibility of the certifier or maker of the statement.

LIGHT + DESIGN: A GUIDE TO DESIGNING QUALITY LIGHTING FOR PEOPLE AND BUILDINGS

Quality of the Visual Environment Committee,
Illuminating Engineering Society of North America

Committee Members:

Chair: Carol Jones
Vice Chair: Leslie North
Secretary: Veda Clark

Subcommittee SC-1: Design Guide
Chair: Peter Ngai
Co Chair: Naomi Miller
Members: Veda Clark
           Dawn DeGrazio
           Carol Jones
           Hayden McKay
           Leslie North
           Yukiko Yoshida

Full Voting Members
Bob Davis
Veda Clark
Dawn DeGrazio
Delores Ginthner
Kevin Houser
Carol Jones
Lorence Leetzow
Terry McGowan
Hayden McKay
Naomi Miller
Guy Newsham
Peter Ngai
Leslie North
Yukiko Yoshida

Advisory Members
John An
Craig Bernecker
Peter Boyce
Wilson Dau
Michelle Eble-Hankins
Peter Hugh
R. Gerald Irvine
JoAnne Lindsley
Martin Moeck
Michael Phillips
Christopher Samuelson
Dyoni Smith
Martyn Timmings
Radosveta Topalova
Jennifer Veitch
Willard Warren
Clarence Waters

The QVE Committee would like to gratefully acknowledge Naomi Miller, FIES, FIALD, for her generous contribution of the original manuscript for LIGHT + DESIGN. We are indebted to her for her gift, which inspired many of us to rise to the task at hand and made it possible to complete this document. The QVE Committee also acknowledges Craig DiLouie as this guide's editor.

TABLE OF CONTENTS

FOREWORD   xiii

PART ONE: LIGHT + HUMAN NEEDS   1

CHAPTER 1. LIGHT + QUALITY   3
  Human Needs Served by Lighting   3
    Task visibility   3
    Task performance   3
    Mood and atmosphere   4
    Visual comfort   4
    Aesthetic judgment   4
    Health, safety and well-being   5
    Social communication   5
  The WHO of Lighting   5
  Light + Human Needs   6
  Light + Economics & Environment   7
  Light + Architecture   8

CHAPTER 2. LIGHT + VISION   9
  Task Visibility   9
  SIDEBAR 2-1. The Human Visual System   11
  Luminance   12
  SIDEBAR 2-2. Measuring Light: Lumens, Lux and Candelas   12
  Contrast   14
  Color Contrast (Color Difference)   18
  Size   18
  Movement   19
  Time   19
  How to Achieve Good Task Visibility   19
  SIDEBAR 2-3. Lamps, Luminaires and Light Distribution   22

CHAPTER 3. LIGHT + VISUAL COMFORT   25
  Visual Comfort   25
  Visual Comfort and Glare   25
  SIDEBAR 3-1. Field of View   26
  SIDEBAR 3-2. Overhead Glare   26
  How to Eliminate Unwanted Glare   27
  Visual Comfort and Flicker/Strobe   30
  How to Minimize Lamp Flicker   31

CHAPTER 4. LIGHT + MODELING — 32
- Modeling of Faces and Objects — 33
- Modeling of Objects — 33
- Modeling of Faces — 34
- SIDEBAR 4-1. Point Source, Linear Source, Area Source — 36
- How to Model Using Light and Shadow — 37

CHAPTER 5. LIGHT + COLOR — 39
- Color Appearance (of Objects, People or Light Sources) — 39
- Color Temperature — 40
- SIDEBAR 5-1. Color Temperature — 41
- Color Rendering — 42
- How to Achieve Good Color Quality — 42
- SIDEBAR 5-2. Color Rendering Index — 43

PART TWO: LIGHT + ECONOMICS & ENVIRONMENT — 47

CHAPTER 6. LIGHT + COST — 49
- Cost, Both Initial and Maintained — 49
- How to Maximize Value on a Given Budget — 49

CHAPTER 7. LIGHT + MAINTENANCE — 51
- Maintenance and Change — 51
- How to Design for Maintenance — 51

CHAPTER 8. LIGHT + ENERGY — 55
- Energy Use — 55
- Design Considerations for Minimizing Energy Use — 55

CHAPTER 9. LIGHT + ENVIRONMENT — 59
- Environmental Considerations — 59
- Efficiency in Whole, Not Part — 59
- Rating Systems Provide Benchmarks — 60
- Lighting's Role in Sustainability — 60
- Cradle to Cradle Products — 60
- Think Globally, Design Locally — 60
- Lamps and the Environment — 60
- How to Support Sustainable Design Goals with Lighting Choices — 60

CHAPTER 10. LIGHT + CONTROL — 63
- Controls for Energy, Flexibility and People — 63
- Controls for Energy — 63
- Controls for Flexibility — 64
- Controls for Occupant Satisfaction — 65
- How to Incorporate Controls into a Project — 65

TABLE OF CONTENTS IX

CHAPTER 11. LIGHT + NIGHT — 69
Emerging Outdoor Lighting Issues — 69
Light Pollution/Light Trespass — 69
Glare and Outdoor Lighting — 69
How to Minimize Light Pollution/Light Trespass and Glare — 69
Peripheral Detection and Outdoor Lighting — 74
SIDEBAR 11-1. Detection and the Driving Task — 74
How to Enhance Peripheral Detection — 75

PART THREE: LIGHT + ARCHITECTURE — 77

CHAPTER 12. LIGHT + ARCHITECTURE — 79
Lighting for Architecture — 79
Appearance of Space and Luminaires — 79
Architectural Integration — 80
Size, Finish, Mounting Height — 81
Location of Luminaires — 81
Furniture Finishes and Sizes — 82
Tradeoffs — 82
How to Improve the Appearance of Space and Luminaires — 82

CHAPTER 13. LIGHT + DISTRIBUTION — 85
Patterns of Light — 85
Light Distribution on Surfaces (Patterns)/Points of Interest — 85
How to Achieve Appropriate Light Patterns on Surfaces and Objects — 89
Sparkle (or Desirable Reflected Highlights) — 90
How to Create Sparkle — 91
Light Distribution on Task Plane (Uniformity) — 92
How to Achieve Good Uniformity on the Task Plane — 93
SIDEBAR 13-1. Importance of Task Plane
 Uniformity in Industrial Applications — 94
Room Surface Brightness (and Surface Characteristics) — 94
SIDEBAR 13-2. Reflectance/Reflectances — 97
How to Achieve Room Surface Brightness — 98
SIDEBAR 13-3. Luminance Versus Exitance — 98

CHAPTER 14. LIGHT + DAYLIGHT — 101
Daylight and View — 101
How to Avoid Glare — 102
How to Optimize Daylighting for Different Building and
 Room Shapes — 105
How to Balance Electric Light with Daylight — 107
How to Save Energy with Daylighting — 108
More Information — 108

CHAPTER 15. LIGHT + CODE — 109
- Code Compliance, Standards and Legal Issues — 109
- United States — 109
  - *Energy codes* — 109
  - *Building codes* — 110
  - *The Energy Policy Act of 2005 (EPAct 2005)* — 111
  - *Americans with Disabilities Act (ADA)* — 111
- Canada — 111
  - *Energy codes* — 111
  - *Building codes* — 112
- Mexico — 112
  - *Energy codes* — 112
  - *Building codes* — 114

CHAPTER 16. LIGHT + SAFETY — 115
- Safety, Security and Emergency Egress — 115
- Safety from Bodily Injury — 115
- How to Light for Safety — 115
- Safety from Hurt and Loss — 116
- How to Light for Security — 116
- Safety during Emergency Egress — 117
- How to Light for Emergency Egress — 117
- SIDEBAR 16-1. Emergency Lighting in the World Trade Center in 1993 — 118

CHAPTER 17. LIGHT + SPECIAL CONSIDERATIONS — 119
- Special Considerations — 119
  - *Conservation of materials (museums)* — 119
  - *Lighting for cameras* — 119
  - *Luminaire noise* — 119

PART FOUR: LIGHT + APPLICATION — 121

CHAPTER 18. A NEW PROJECT: COLLECTING INFORMATION — 123
- Client Goals and User Needs — 123
- The WHO of Lighting Design — 123
- Human Needs — 123
- Economics, Energy and the Environment — 124
- Architecture and Other Building-Related Issues — 125

CHAPTER 19. LIGHTING DESIGN GUIDE — 127
- Lighting Design Guide — 127
- How to Use the Lighting Design Guide — 128
- Summary of Differences between 2000 and Updated Lighting Design Guide — 129
- Lighting Design Guide — 130

TABLE OF CONTENTS	XI

CHAPTER 20. APPLICATIONS GUIDE — 133

Applications Guide — 133
#1 Daylighted Classroom — 134
   Context and Objectives — 134
   Issues and Guidelines — 134
#2 Open Office (Direct/Indirect Luminaires and Task Lighting) — 136
   Context and Objectives — 136
   Issues and Guidelines — 136
#3 Open Office (Advanced Controls and Personal Dimming) — 138
   Context and Objectives — 138
   Issues and Guidelines — 138
#4 Open Office (Direct Lighting Using Recessed Parabolic Luminaires) — 140
   Context and Objectives — 140
   Issues and Guidelines — 140
#5 Industrial Assembly Area/Manufacturing — 141
   Context and Objectives — 141
   Issues and Guidelines — 141
#6 Industrial Warehouse — 144
   Context and Objectives — 144
   Issues and Guidelines — 144
#7 Bank Lobby with Teller Line — 146
   Context and Objectives — 146
   Issues and Guidelines — 146
#8 Big Box Discount Store — 147
   Context and Objectives — 147
   Issues and Guidelines — 147
#9 Retail Clothing Space — 148
   Context and Objectives — 148
   Issues and Guidelines — 148
#10 Supermarket — 149
   Context and Objectives — 149
   Issues and Guidelines — 149
#11 Mall Chain Store — 151
   Context and Objectives — 151
   Issues and Guidelines — 151
#12 Banking Office Entrance and Parking Lot — 153
   Context and Objectives — 153
   Issues and Guidelines — 153
#13 Courtroom — 155
   Context and Objectives — 155
   Issues and Guidelines — 155
#14 Audience Chamber for Theatre — 157
   Context and Objectives — 157
   Issues and Guidelines — 157

APPENDIX I. GLOSSARY OF TERMS — 159

APPENDIX II. ENDNOTES — 163

INDEX — 171

Minneapolis Convention Center, Lighting: Schuler Shook Theatre Planners Lighting Designers; Photographer: Myunghwan Cho

FOREWORD

Our relationship to our surroundings is directly related to the quality of the lighted environment. When the light is right, we see more deeply into the world. Quality lighting enhances our ability to see and interpret the world around us, supporting our sense of well-being and improving our capability to communicate with each other.

Because lighting is vital for vision, activity and perception, it is critical to provide a quality lighted environment for people who use lighting systems for work or leisure.

LIGHT + DESIGN was developed to introduce architects, lighting designers, design engineers, interior designers and other lighting professionals to the principles of quality lighting design. These principles, related to visual performance, energy and economics, and aesthetics, can be applied to a wide range of interior and exterior spaces to aid designers in providing high-quality lighting to their projects.

The text is divided into four parts. Part I: Light + Human Needs (Chapters 1-5) describes the principles and technical background involved in designing lighting that satisfies vision and visual comfort needs. Part II: Light + Economics & Environment (Chapters 6-11) deals with economic and environmental factors related to quality design. Part III: Light + Architecture (Chapters 12-17) deals with lighting the built environment. And Part IV: Light + Application (Chapters 18-20) presents the IESNA Design Guide, introduced in the ninth edition of the IESNA Lighting Handbook, updated and excerpted for this guide; an applications guide; and a list of questions that provide a checklist to identify issues critical to your specific project. Appendices, including a glossary and extensive endnotes, are also provided.

LIGHT + DESIGN is illustrated with sketches and photographs to enhance understanding of key principles and provide examples of applications. The illustrations portray many typical applications as well as more glamorous installations.

This guide focuses on design principles but defines key technical terms and includes technical background in the form of sidebars to aid understanding. Sidebars are identified as "basic" or "advanced" according to the level of explanation and technical knowledge. A glossary provides more rigorous definitions of lighting terms and metrics.

In addition, this text references other technical papers and IES publications that can provide additional research background, detail or specific design guidance. These documents, referenced in the endnotes, provide an extensive background of technical lighting knowledge and access to the latest research influencing our understanding of light.

Quality lighting design distinguishes the designer, provides full value to the client, and supports the design intent. LIGHT + DESIGN will introduce you to the principles and background involved in achieving this level of lighting for people and buildings.

Carol Jones, Chair, IES QVE Committee
Naomi Miller, FIES, FIALD, Vice Chair,
Design Guide Task Force Subcommittee

PART ONE: LIGHT + HUMAN NEEDS

To love beauty is to see light.

– Victor Hugo

LIGHT + QUALITY

Following the light of the sun, we left the Old World. – Christopher Columbus

Vision and light enable us to understand the physical universe. Light is the portal through which the eye, and thereby the human mind, interacts with the world, and is therefore capable of creating worlds of perception. Based on this understanding of light as a medium, the composition of light in a space can inspire a wide range of psychological and even physiological responses in humans.

Throughout the ages, people spent most of their time outdoors and relied on sunlight, moonlight, starlight, firelight. During the day, people had little control over the lighting that they were given from nature, but it was dynamic. The moving sun continually created subtle changes in the landscape, revealing and obscuring texture and detail through light and shadow, color and brightness.

Today, people in industrialized nations spend most of their lives indoors and rely on electric light as their medium to interpret architectural environments. Electric light enables designers to control the medium, use it for communication, and thereby inspire different perceptions and achieve a range of functional and aesthetic goals.

Good lighting provides sufficient illumination for people to see and perform visual tasks; it can also draw attention, influence social interaction, foster mood and atmosphere, beautify space and architecture, promote safety and security, increase comfort and contribute to task performance. Poor lighting can have the opposite effects.

Human Needs Served by Lighting

Task visibility
Task visibility is essential to lighting design; lighting exists to enable vision. Recognition of this fact led to an emphasis on visibility above all other lighting design goals in the past, resulting in a high level of understanding of visibility and its importance.

Task performance
Task performance involves occupants interacting with objects in a space, whether it is a person washing his or her

Figure 1-1. *Human needs served by lighting.*

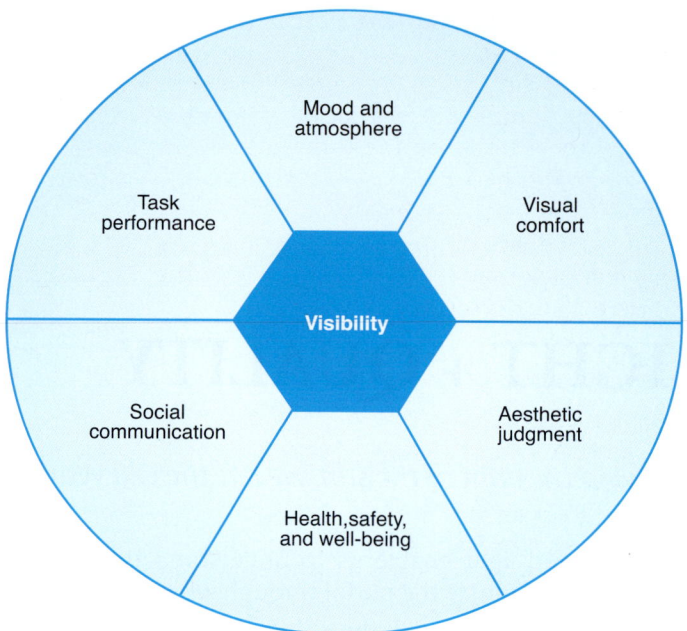

hands, reading numbers posted in a corridor, analyzing the details in an etching, or any other task. While lighting is a critical component of visual performance, visual performance in turn is only one of several influences on task performance along with training, motivation, motor skills and other factors.[2, 3]

Mood and atmosphere

These concepts reflect emotional response to a luminous environment.[3] Lighting design can have a direct impact on many aspects of human experience, including relaxation, stimulation, preference and sense of safety. These mood states in turn may influence task performance.[4]

Visual comfort

Visual comfort can affect task performance, health and safety, and mood and atmosphere. Lighting conditions that cause visual discomfort can lead to headaches and eyestrain. Visual discomfort and its effects are often dependent on the context of the application. For example, office workers often find themselves more fatigued under a glaring lighting installation, while dancers in a discotheque might find glare an enhancement to the dancing experience.

Aesthetic judgment

Aesthetic judgment in lighting design is distinguishable from emotional response. Humans have an inherent need to make sense of what they see; in any particular setting, therefore, visual information must be clearly illuminated and comprehensible. Lighting is often used to communicate meaning and reinforce the rhythmic patterns of architecture. It can also enhance color or emphasize a visual hierarchy of focal points in a space. Various models have been developed in an attempt to quantify aesthetic judgments. One uses four dimensions of appraisal: coherence, legibility, mystery and complexity.[5, 6] Another uses visual interest and visual lightness (room surface brightness).[7, 8, 9] Both theories conclude that humans do not prefer uniform, homogenous lighting. However, extremes in one of these factors may reduce the impact of another; for example, a scene that is highly complex may result in lower coherence. In this case, lighting may contribute to visual clutter.

Health, safety and well-being

These are essential, but often overlooked, aspects of lighting. There are numerous ways lighting relates to this aspect of human needs. For example, magnetic ballast operation may produce a sensation of lamp flicker among some people, leading to eyestrain and acting as a trigger for headaches or migraines.[10, 11] Safety is a paramount concern, requiring effective emergency lighting and also appropriate visibility for features such as curbs, stair edges, train platforms, roadway intersections and labels of critical chemicals and pharmaceuticals. Light processed by the eyes has a direct effect on alertness and the state of the body's circadian system (i.e., the sleep-wake cycle). Exposure to light at critical times of night may suppress the release of the hormone melatonin in the brain; recent research suggests that disruption of the circadian system may have long-term consequences for different types of cancer.[12] This is a new, rapidly evolving field of research. Finally, visually impaired people may have special lighting needs. For example, some low-vision people require much higher light levels at the task, while other groups, such as those with albinism, prefer subdued lighting because they are extremely light-sensitive.[13]

Social communication

Social communication is subject to a commanding influence from lighting design. The quality of illumination is particularly important when people speak to each other face to face, close-up as in an intimate restaurant, at a medium distance as in a conference room, and at a distance as in a house of worship. Facial recognition is a critical element in security lighting, which must provide a sufficient quantity of light to make the face visible. The modeling of facial features created by the pattern of light and shadow is also important. In addition, light is essential for nonverbal communication. Cues, gestures and expressions can be missed or misunderstood under poor lighting.

The WHO of Lighting

Because the overall purpose of lighting is to serve human needs, designers must approach every project not only from the perspective of the client, but also from the perspective of the ultimate users of the lighted space. The first step to a successful lighting design, therefore, is to assess the purpose of the space and user needs along with overall client objectives:

Figure 1-2. *Perspectives on lighting goals can vary across an organization. (Drawing courtesy of Yukiko Yoshida from original by Naomi Miller.)*

Who are these users? What are they expecting to see? What are their needs?

After assessing lighting needs, the designer then ranks and correlates architectural objectives along with economic and environmental considerations.

Who will own it? Who will install it? Who will pay for its operation? Who will maintain it?

Ultimately, the sum of these factors must be translated into a workable design and functional installation. But understanding the *who* of a project is a critical first step.

The result, properly researched, conceived and executed, contains the elements of a quality lighting design—addressing human needs, economic and environmental considerations, and architectural priorities.

Light + Human Needs

Quality lighting design:

- Provides sufficient light (i.e., *illuminances*—see Chapter 2 for definition and discussion) where primary visual tasks are performed.
- Provides proper *luminances* (which can affect user perception of brightness—see Chapter 2 for definition and discussion), delivers uniformity of illumination whenever appropriate, and does not produce glare that can impede task performance.
- Properly models faces and objects using light and shadow.
- Uses optimal *color temperature* and *color rendering* (see Chapter 5 for definition and discussion) for light sources to both enrich and increase the accuracy of the visual experience, whenever color quality is important.

CHAPTER 1: LIGHT + QUALITY

Figure 1-3. *Quality lighting enhances architecture. Dulce Health Care Facility. (lighting design: Don Gallegos, IES, WHPacific, Inc., Albuquerque, NM; photographer: Kirk Gittins)*

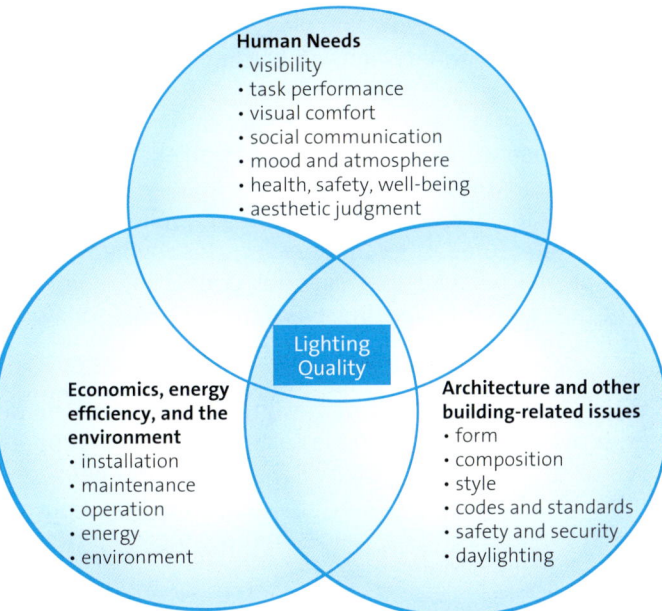

Figure 1-4. *Lighting quality: The intersection of human needs, architecture, economics, energy efficiency, and environment.*

Light + Economics & Environment

A quality lighting design:

- Is easy to maintain so that the design intent will continue to be met on an ongoing basis.
- Provides a lower ownership cost of the lighting system by choosing energy-efficient components and *luminaires* (see Sidebar 2-3 for definition) with the minimum standard being the applicable energy code.
- Is environmentally sensitive in its energy consumption, use and disposal of materials, and conservation of natural resources.
- Utilizes lighting control dimming and switching strategies to increase flexibility and maximize energy savings and occupant satisfaction, whenever appropriate.
- Minimizes skyglow and eliminates light trespass and glare in exterior lighting installations.

Light + Architecture

Quality lighting design:

- Integrates properly with architecture either by using luminaires that stand out, blend in or completely disappear into the architecture.
- Uses lighting emphasis or variation to highlight points of interest or create a visual hierarchy.
- Illuminates walls and ceilings to improve user satisfaction and comfort.
- Uses daylight, whenever appropriate.
- Complies with all applicable building electrical and energy codes.
- Makes human safety and security top priorities.

Note that while higher budgets increase options and capabilities, designers with knowledge, creativity and experience can provide quality lighting to clients even given constrained budgets.

Not all projects are appropriate applications of all quality lighting approaches. Quality lighting is, to an extent, a moving target. In other words, what constitutes quality lighting for a given application always depends on that application's common and unique needs. In some cases, this involves achieving an optimal balance that satisfies conflicting goals. The thought that goes into lighting every space should follow the above principles, but be fresh, reflecting current best practice and technology.

Ultimately, quality lighting design distinguishes the designer, provides maximum value to the client, and supports the overall design intent. Quality lighting not only helps people do things, it can promote occupant satisfaction while doing this; it can differentiate spaces that are merely functional from those which people enjoy.

2

LIGHT + VISION

Task Visibility

The foundation of lighting design is ensuring that people have enough light to safely, efficiently and accurately perform predominant visual tasks.

First, a definition of *visual task*: The object or area that people need to see in order to gather information, interact with, or manipulate objects.

Threading a needle is a visual task. The person needs to see the tip of the thread and the hole in the needle. He or she also needs depth perception cues to know when the thread is nearing the hole in the needle.

Reading the printed words on a glossy magazine page is a visual task, as is reading numbers in a hotel corridor and spotting dirt on the floor of a corridor that is being cleaned. Seeing the red or green characters in an exit sign in an emergency is a critical visual task.

Seeing faces is another, whether it is in a guard's station, a house of worship or in a courtroom where the guilty and innocent are identified.

A primary objective of quality lighting design is to assess the visual task needs in a space and ensure that the lighting system provides sufficient *illuminance*, or light level, for users to be able to perform visual tasks safely, efficiently and accurately. Achieving proper task visibility requires knowledge of the three-dimensional relationship between the user, the task and the light source—or the *source/task/eye geometry*—and also the space conditions that may affect these relationships.

There are many factors that affect our ability to see, the six most important being Luminance, Contrast, Color Difference (sometimes called Color Contrast), Size, Movement and Time. Fundamentally, for each task:

Is there enough contrast and color difference between the task detail and its background for it to be visible?

What is the size of the visual task, and, because this also affects size, how far from the task is the viewer?

Is the task moving, is this movement predictable, and at what speed?

How much time does the occupant have to complete the task?

These factors, to an extent, are interlinked. For example, one can often read smaller print if the light level is increased,

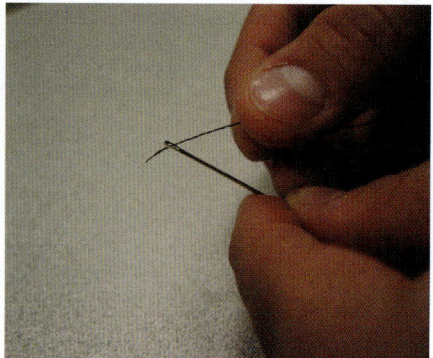

Figure 2-1. *Threading a needle.*

Figure 2-2. *Reading a magazine.*

Figure 2-3. *Cleaning dirt from a floor.*

Figure 2-4. *Seeing an exit sign in a corridor.*

or if one has more time to study the page. The designer can influence these factors during the design process.

Related issues include: How important is speed of task completion? How important is accuracy? Are there special safety issues?

This chapter describes the factors that affect task visibility. These factors influence designer choices of how much light must be present at the visual task and, in turn, the output of the lighting system to be designed.

SIDEBAR 2-1. The Human Visual System

The human visual system involves the eye and brain. The eye receives visual information and passes it on to the brain for higher level visual processing.

The important optical components of the eye are the curved cornea, which is responsible for most of the optical power of the eye; the pupil formed by the iris, which opens in dim light and closes in bright light; and the lens, which bulges to focus on near objects and flattens to focus on objects far away.

The optics of the eye are designed to form an image of the world on the photosensitive surface of its retina. The photoreceptors of the retina absorb the incident photons of light and convert them into electrical and electro-chemical signals. These signals are processed in the retina and then transmitted to the visual cortex of the brain, where the outside world is reassembled in the process of vision.

The retina contains two types of photoreceptors, called rods and cones according to their shape.

The 120 million rods serve as a night retina, although rods also participate in collecting visual image information even under higher light conditions. Rods are characterized by a high sensitivity to light but without an ability to see detail or color.

The eight million cones serve as a day retina, characterized by an ability to see fine detail and color. For most interior lighting, vision is dominated by cones. Cone vision is also called photopic vision. For most exterior lighting, vision is influenced by both rods and cones. With both rods and cones active, this is called mesopic vision. Electric lighting generally produces enough light even at relatively small levels that vision is rarely, if ever, governed by the rods alone. When the rods alone are active, this is called scotopic vision.

The rods are distributed widely across the retina with the exception of a small area at the center of the retina called the fovea. The cones are also distributed across the retina but they are strongly concentrated in the fovea.

The distribution of rods and cones across the retina means that visual abilities are usually considered in two separate areas: 1) On-axis, when the retinal image of the object is formed on the fovea, and 2) off-axis, when it is not.

When the cones are operating, visual acuity, contrast sensitivity, color discrimination and many other visual capabilities are at a maximum on-axis and decline markedly off-axis. This explains the usual operation of the visual system where something of interest is first detected off-axis (i.e., "seen out of the corner of your eye") and then the viewer shifts his or her focus towards the object so that it can be viewed on-axis and recognized.

The retina also contains ganglion cells, some of which perform pre-processing of signals that get sent to the visual cortex of the brain, and others that send signals to the hormone centers of the brain and may be important for human health and immune system function.[14]

Visual systems are remarkably resilient; the human eye can see under full moonlight (approximately 0.1 lux or 0.01fc)* and under bright noon daylight (approximately 100,000 lux or 10,000 fc)**, but not at the same time. It takes time for human visual systems to adapt to a change in environmental brightness. How long it takes depends on how big a step in luminance occurs and what photoreceptors are expected to be operating.

At one extreme, changes of about 1000:1 in daytime conditions can be adapted to in less than a second, but adapting from bright sunlight to a darkened theatre can take as long as 30 minutes.

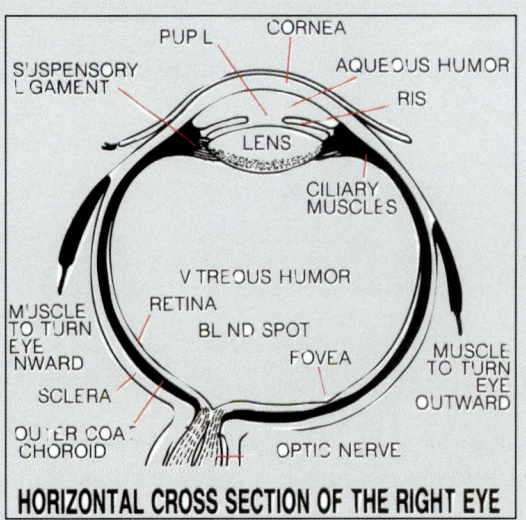

Figure 2-5. *A cross-section through the human eye, showing the retina where the visual field is projected.*

---

* (0.006 cd/m² which is approximately 0.1 lux falling on a surface of 18% reflectance)
** (6,000 cd/m², which is approximately 100,000 lux falling on a surface of 18% reflectance)

Luminance

Luminance is a measurable quantity that results when light flux interacts with a surface by passing through it or being reflected from it, towards the eye of the observer.

Increasing the amount of light striking a task will increase its luminance. Typically, this will result in greater task visibility, except in the case of reflected glare that produces a *veiling reflection*.

Note that luminance should not be confused with *brightness*. Brightness is subjective, based on perception. Luminance of an area can be quantifiably measured. Brightness of an area, on the other hand, depends not only on its luminance, but also its size and the light levels to which the viewer is adapted.

For example, if you have a white page and a gray page of paper on your desk, and the same amount of light flux (lumens) is falling on both pages, the white page will be higher in luminance and it will appear brighter to you than the gray page.

Or suppose a flashlight is shining in one room with the lights on and in another with the lights off. The flashlight would appear brighter in the room with the lights off than the room with the lights on because the viewer's eyes are adapted to a lower luminance. The lighter surrounds make the flashlight appear less bright, even though the measured luminance of the flashlight is the same under both conditions.

SIDEBAR 2-2. Measuring Light: Lumens, Lux and Candelas

Consider a basic scenario consisting of a light source, an object and a person:

Light Exits the Light Source: The light source produces visible light, termed light output or *luminous flux*, which is measured in lumens. Light energy expended over time is measured as *lumen-hours*.

While the lumen metric expresses luminous energy being expended, it does not provide information about how the light is distributed. Does the luminaire emit light in all directions, in a concentrated beam, or in a particular pattern of intensity?

The candela metric is used to measure the intensity of light emitted by a light source or luminaire in a given direction. A beam of light, for example, may be described as having intensity in candelas, formerly called *candlepower*. Graphs depicting intensity and direction—called

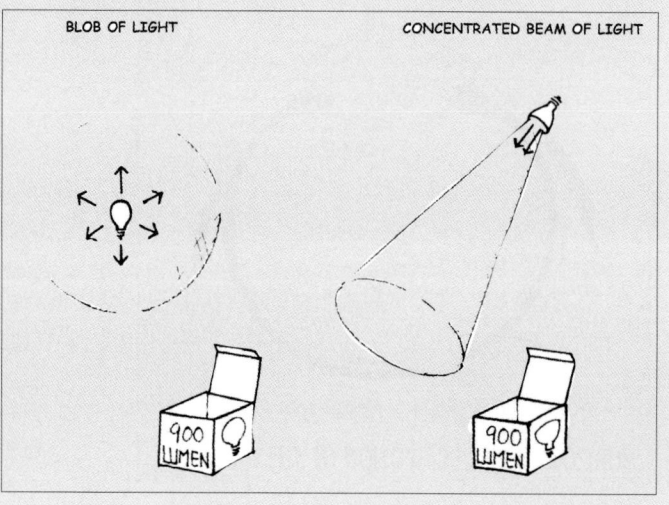

Figure 2-6. *A lamp or luminaire emits luminous flux, measured in lumens. This metric, however, only reveals gross light output, not the direction or intensity of the emitted light. (Drawing courtesy of Yukiko Yoshida.)*

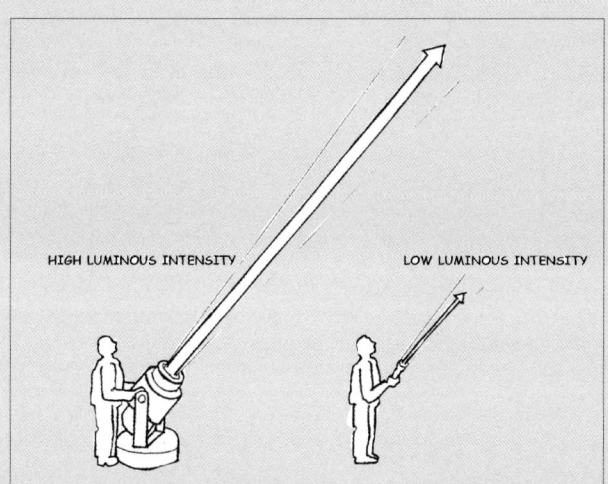

Figure 2-7. *Luminous intensity, measured in candelas, provides information on intensity of emitted light in a given direction. (Drawing courtesy of Yukiko Yoshida.)*

Figure 2-8. *Illuminance levels. (Drawing courtesy of Yukiko Yoshida.)*

candela distribution curves, typically found in the photometric report for luminaires and some directional lamps—can be used to describe a lamp or luminaire's intensity distribution.

Light Falls on a Surface: The light strikes the object, where it is measured in *lux* (metric units) or *footcandles* (English units), as *illuminance*. Illuminance expresses the density of light falling on a surface. A lux (lx) is one lumen falling on a surface that is one square meter in area. A footcandle (fc) is one lumen (lm) falling on a surface that is one square foot in area. One footcandle equals 10.76 lux, although it is sometimes approximated to 10 lux.

The human eye can adjust to a wide range of illuminances, including about 100,000 lx (10,000 fc) on some sunny days to about 0.1 lux (0.01 fc) under full moonlight. Parking lot lighting produces about 10 lux (1 fc) on the ground. Office and classroom lighting often measure between 300 and 500 lux (30 and 50 fc) on the desktop.

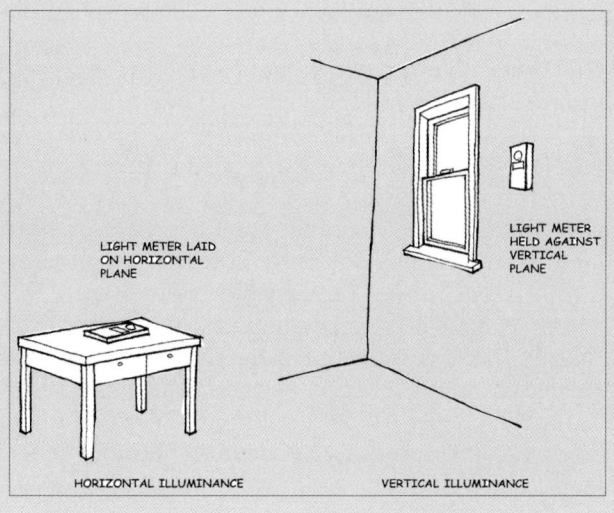

Figure 2-9. *Illuminance measured on a horizontal or vertical plane. (Drawing courtesy of Yukiko Yoshida.)*

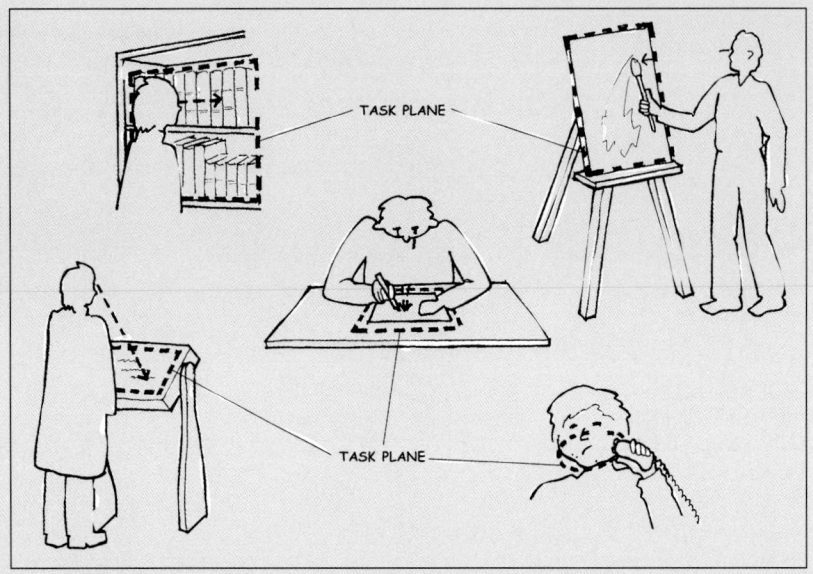

Figure 2-10. *Examples of different task planes. (Drawing courtesy of Yukiko Yoshida.)*

Illuminances are often designed to be uniform within a defined area across the space's *task plane*, or *workplane*. This is an imaginary plane in a space on which visual tasks are performed. The plane may be horizontal, vertical or slanted; a given space may have several task planes.

Light is Reflected to the Human Eye: Based on its optical characteristics, the object absorbs some wavelengths of light and reflects others. The light radiated towards a specific direction from the object is *luminance*, measured in *candelas per square meter* (candelas/m2).* The amount of luminance produced depends on the direction and intensity of incident light on a surface, the surface's reflective properties, and the observer's viewing direction. Luminance is subjectively perceived by the viewer as *brightness*.

---

* The obsolete English system unit, called footlamberts (fL), can still be found in some luminaire photometric reports. To convert fL to cd/m² (on a perfectly diffuse surface only), multiply by 3.4.

Contrast

Every visual task has some combination of light and dark areas that the human visual system must discern in order to "see" it. This is called *contrast*.

It could be a black character printed on a white page, or the shadow line of a nose against a cheek in facial features, or a shadowy area of dirt on a floor.

Contrast may be affected by source/task/eye geometry, disability glare, shadows and the surface characteristics of objects being viewed.

Source/Task/Eye Geometry: Each visual task also has an inherent three-dimensional relationship between the viewer's eye location, the visual task and the location of the light source. Changing any one of the factors in the source/task/eye geometry changes how visible the task is.

For example, consider reading a glossy magazine. At one viewing angle, there is reflected glare from a light source which produces a veiling reflection on the page, obscuring its content from the viewer. To correct the problem and enable seeing the magazine content, the viewer could change his or her location, the viewer could move the magazine, or the viewer could move the light source.

CHAPTER 2: LIGHT + VISION

Contrast can be high or low.
Contrast can be high or low.
Contrast can be high or low.
Contrast can be high or low.
Contrast can be high or low.
Contrast can be high or low.

Figure 2-11. *Characters printed in different shades of gray against a white background illustrating degrees of contrast. (courtesy of Holophane, an Acuity Brands Co.)*

Figure 2-12. *People see facial features through luminance contrast.*

Disability Glare: Disability glare is created by light being scattered in the eye, thereby casting a luminous veil across the retinal image, as though the viewer were suddenly looking at a scene through a sheer curtain or smeared eyeglasses. The effect of this veil is to reduce the contrast of the retinal image and de-saturate any colors.

The eyes of older individuals have more opacities and elements that fog the visual image in them, and so are prone to scatter more light. As a result, these people are more likely to experience disability glare.

Disability glare is a common problem for drivers on roads at night when driving toward the bright headlights of an oncoming vehicle.

A source of bright light is more likely to be disabling if it is close to the direction of view.

A good test for disability glare is to shield the glare source with one's hand; if visibility improves once the glare source is blocked, disability glare exists.

Figure 2-13. *Disability glare producing scatter and washing out of image. (Photo courtesy of International Dark Sky Association (IDA).)*

Figure 2-14. *Same view without disability glare. (Photo courtesy of International Dark Sky Association (IDA).)*

Figure 2-15. *Poor contrast on glossy magazine page.*

Figure 2-16. *Sketch of three-dimensional relationship (Drawing courtesy of Yukiko Yoshida from original by Naomi Miller.)*

Shadows: affect task visibility by reducing contrast. Consider a traveler, flying on an airplane at night, who turns on the overhead "reading" light to write a letter. The strong shadows created by his or her hand on the paper can make it impossible to see details such as where the pen is positioned on the paper. Hand or body shadows, where a person's own body blocks the light required for that person to see, can be detrimental and even unsafe for performing visual tasks.

In some cases, however, shadows can enhance a task's visibility by increasing its contrast, as in the case of using the graduated scale of a micrometer. The lines inscribed in metal are difficult to read unless the shadow makes the line appear dark.

Figure 2-17. *Improved contrast by changing source/task/eye geometry.*

Figure 2-18. *Sketch of improved three-dimensional relationship. (Drawing courtesy of Yukiko Yoshida from original by Naomi Miller.)*

Figure 2-19. *Poor contrast on sewing task.*

Intrinsic Material Considerations: The ability to see important details is dependent on the characteristics of the material being viewed, including how shiny or matte a finish is; its color, both hue and consistency; how smooth or grainy it is; etc. Workers often need to see surface characteristics as a part of manufacturing processes. For example, if working in an automobile paint line, it is important to see drips or runs that may be more shiny or thicker than the surrounding paint.

Once those characteristics are understood, it is easier to determine where the light source should be located, what color features it should have, whether the light source should be focused or diffuse, and so on. The position of the light source relative to the visual task and the worker's line of sight can enhance or hinder the visibility of the surface features.

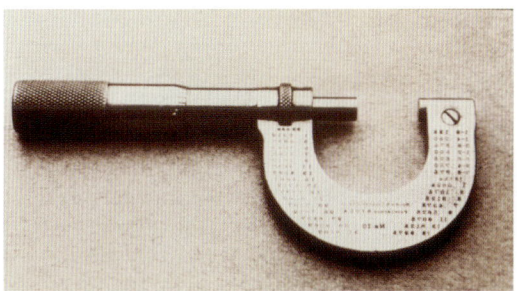 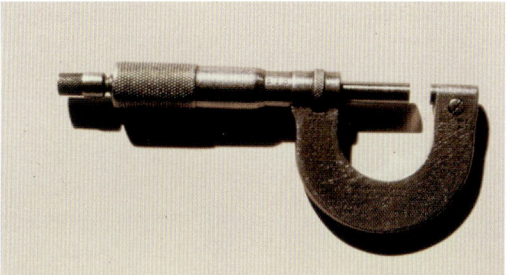

Figure 2-20. *(left);* **Figure 2-21.** *(right)*
The etched graduations on this micrometer are easier to see in Figure 2-20 (left) than in Figure 2-21 (right). In Figure 2-21, the directional lighting does a better job than the diffuse lighting to help create shadows in the grooves that enhance the contrast of the mark against the steel shaft.

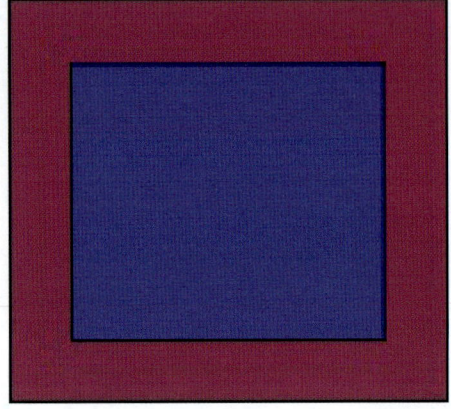

Figure 2-22. *The blue and purple colors appear different because of color contrast, despite equal luminance. (Original graphic courtesy of Naomi Miller.)*

Color Contrast (Color Difference)

The human visual system uses the difference in color between two adjacent patches to discriminate objects, especially when there is little luminance contrast between the two areas. That is, the two patches will look different even though they reflect the same amount of light toward the viewer's eye.

Assuming equal luminances from a sign's lettering and background, for example, yellow lettering on a blue background is more visible than orange lettering on a brown background.

However, the luminance and color difference of this sign could be altered based on the color qualities of the light source selected to illuminate it. Color is described in greater detail in Chapter 5.

Size

Generally, the larger the visual task, the easier it is to see. For example:

This half of this sentence is easier to read

than this half of this sentence.

The apparent size of the visual task depends on the actual (physical) size of the task, the viewer's distance from it and the angle of view.

For example, 12-point type is likely easy to read at arm's length, but if this guide were located 6 meters (20 feet) away, it would become impossibly small to read. Conversely, the fine print in a contract can be made readable if the viewer moves his or her head closer to the paper and is able to maintain focus, effectively increasing the size of the task.

Similarly, faces across a large room are difficult to recognize until the individuals have moved closer to the viewer, effectively increasing the size of their facial features.

Movement

Movement is a related factor to task visibility because it affects detection.

The eye can be trained to track objects such as a tennis ball or components for inspection moving along a conveyer belt, but the human visual system is simply not an expert at spotting or tracking moving objects. If an object is moving, particularly if its movement is unpredictable, then more contrast or luminance, or a larger size, is required for people to see it and determine its position with accuracy.

Movement is a critical reason why driving a vehicle is a difficult visual task. It is therefore sometimes important to render objects (that are moving relative to the driver) more visible through nighttime electric lighting.

However, note that there are times when movement can aid detection. For example, an object moving in one's peripheral vision can enhance detection. When scanning for friends in a crowd, it is easier to spot them if they are waving. Similarly, the motion of an animal in the woods will draw attention, although one cannot see enough detail to identify the animal until one focus his or her central vision (i.e., foveal vision) in that direction.

Time

If there is time pressure, it is more difficult to complete a visual task. For example, reading this guide under very dim lighting conditions may be possible, given enough time, but taking an exam under very dim lighting conditions would likely result in a dismal grade.

If seeing details quickly is critical, then task contrast, luminance and size must be sufficient to provide visibility without forcing the viewer to take extra time for the seeing process.

How to Achieve Good Task Visibility

- √ Many visual tasks are not very demanding and require no special considerations. However, if the task is small, of very low contrast, or is covered with highly reflective glass or plastic, the designer should first consider whether it is possible to change the task itself to improve contrast, size or luminance in order to improve visibility without increasing illuminances.

- √ If the task cannot be changed, then select illuminance, light distribution and source/task/eye geometry that will make the task easy to perform. Due to the wide variety of tasks performed in spaces, it is outside the scope of this guide to address all of them. For more information about task visibility, see Chapter 26 of *The Handbook of Human Factors and Ergonomics* by P.R. Boyce.[15]

- √ Select target illuminance values based on task, visual ability of the user, psychological needs and past experience. For a summary of target illuminances, see the IESNA Design Guide excerpt tables in Chapter 19. (The complete tables can be found in the current edition of the IESNA *Lighting Handbook* or the IESNA *Ready Reference Guide*.) These illuminance values have been selected through a consensus process involving experienced designers and engineers, but should be modified based on the conditions of the specific user, task and space.

√ In some cases, the minimum recommended IES illuminances are appropriate for visual task performance, but may fail to meet occupants' other visual needs. The human visual system is remarkably resilient and able to "see" under even dismal conditions. For office lighting tasks, it is difficult to detect significant differences in task performance between 300 lx (30 fc) of illuminance and 500 lx (50 fc) of illuminance. This does not mean the designer should always provide the lower light conditions for office workers; lighting has many psychological and physiological impacts beyond mere task visibility. Someone once said: "You can get clean by taking a cold shower instead of a hot shower, but *would you want to?*" Similarly, one can perform certain visual work under gloomy conditions, but would you *want* to spend eight hours a day there?

√ When selecting target illuminance values for workspaces, high illuminances need not be applied everywhere in the space, and in many cases should not be.

- Use localized (task) lighting to provide high illuminances at task locations, but avoid task illluminances that exceed three times the ambient illuminance so as to avoid noticeably dim areas—i.e., uncomfortable luminance ratios.[16]
- In some applications, however, it is appropriate to achieve higher luminance ratios to achieve a certain effect. For example, high luminance ratios can help attract shopper attention in a retail space.
- Let the less-concentrated general lighting in the room to provide the work of lighting walls, faces, floors and less critical tasks.

√ There are many ways to reduce veiling reflection problems for critical visual tasks.

- In some cases, the geometry of the task or lighting can be changed to avoid a problematic reflection angle.
- Sometimes, the task can be changed for use with matte finish materials, which produce no veiling reflection, as opposed to shiny finishes.

CHAPTER 2: LIGHT + VISION

- If the luminaire must be located in a position that is likely to cause problems, such as directly in front of the viewer in an office workstation, it can be specified with a light distribution so that a minimum of its light is emitted at problem angles. In office lighting, task lighting located under overhead cabinets is often specified with a batwing-distribution lens to accomplish this. (See Section 3.0 of IESNA RP-1-2004 *Recommended Practice for Office Lighting*.)[17]
- To avoid reflected window glare in computer screens, work with space planners and users so that the screens do not face windows.
- To avoid glare from luminaires on computer screens, select lighting systems that will minimize distracting reflections, based on the angle of the screen and the locations where luminaires are likely to cause reflection.
- Computer screens that are less susceptible to reflection, such as flat LCD screens and/or screens with anti-reflection coatings, can mitigate distracting reflections. (For more information about choosing computer screens, see P.41 of IESNA RP-1-2004.)

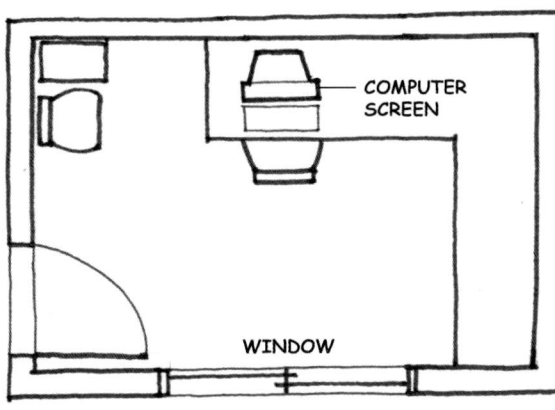

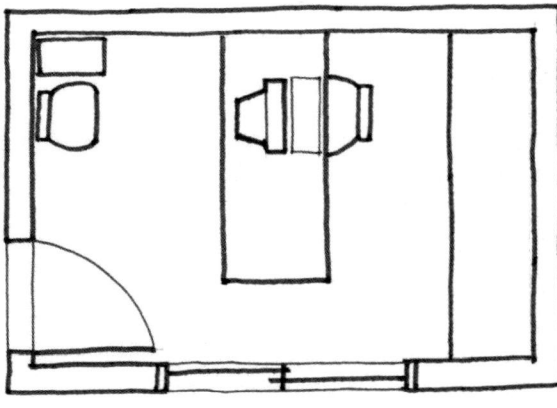

Figure 2-23. *Plan of office, showing orientation of computer screen with respect to windows, to minimize reflected glare. (Drawing courtesy of Yukiko Yoshida.)*

SIDEBAR 2-3. Lamps, Luminaires and Light Distribution

A *lamp* is correct usage for what is sometimes called a light bulb. A lamp is the actual light-producing element, or light source, in a lighting system.

A *luminaire* is correct usage for what is commonly termed a *lighting fixture* in the electrical industry. A luminaire, in its simplest form, includes the light source, a ballast (if applicable), wiring and components that provide electrical connections, the housing that protects these components, and an optical system that directs light emission in desired direction(s) and pattern.

Lamps and luminaires emit light in various three-dimensional patterns.

A basic incandescent lamp emits light almost equally in all directions, while a halogen or LED flashlight emits a narrow cone of light.

Similarly, luminaires can use reflectors and lenses to collect light from the light source and direct it into wide, medium or narrow beams, and upward, downward, outward or a combination of directions.

These capabilities are design tools. By understanding the distribution characteristics of lamps and luminaires, designers can identify luminaires that are suitable for different application needs. For example, the designer may select a direct-indirect luminaire that directs a wide distribution of light towards the ceiling while directing some light directly downward towards the workplane, or an asymmetrical linear uplight to "wash" a distinguished wood ceiling in a hotel lobby.

This information is captured during the photometering process, in which a lamp or luminaire is mounted on a device that collects data about where it emits light and at what luminous intensity (candelas) at these angles. This results in a photometric report describing lamp or luminaire performance as a table of numbers that can be translated into a polar graph that illustrates light distribution.

Figure 2-24. *Examples of luminaires. (Photo collage courtesy of Yukiko Yoshida.)*

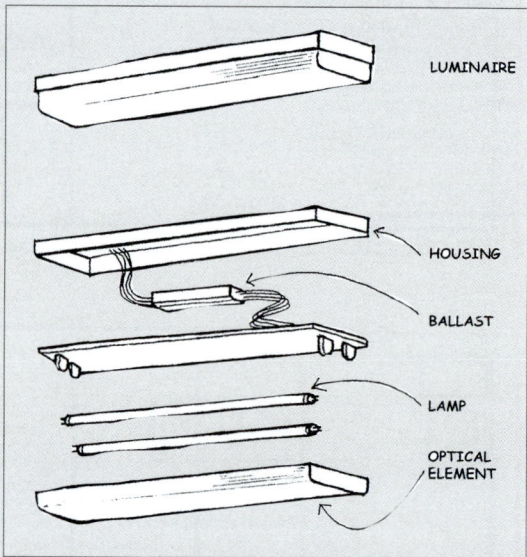

Figure 2-25. *Components of a typical luminaire. (Drawing courtesy of Yukiko Yoshida.)*

CHAPTER 2: LIGHT + VISION

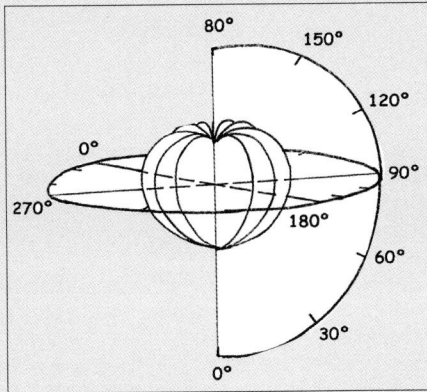

Figure 2-26. *Photometric light distribution from a typical incandescent lamp. Imagine the light source located at the intersection of the three axes.*

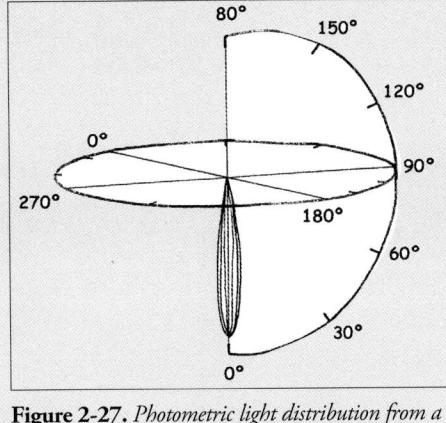

Figure 2-27. *Photometric light distribution from a halogen or LED flashlight that emits a narrow cone of light. Imagine the flashlight located at the intersection of the three axes.*

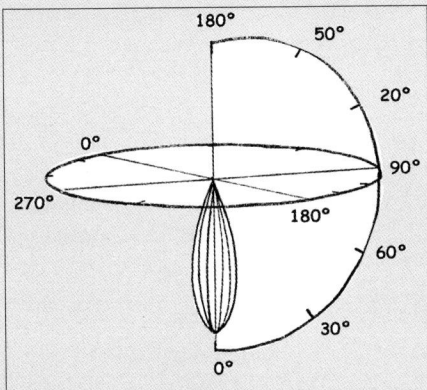

Figure 2-28. *Photometric light distribution from a narrow focus luminaire used over a jewelry counter.*

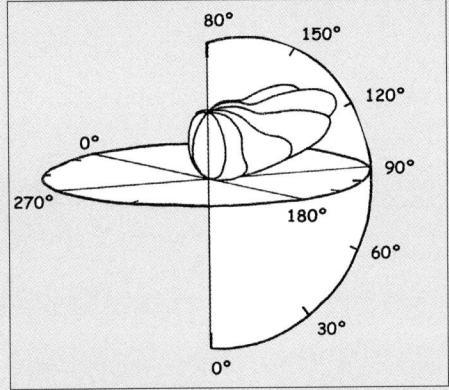

Figure 2-29. *Photometric light distribution from an asymmetrical linear uplight used to wash a wood ceiling in a hotel lobby.*

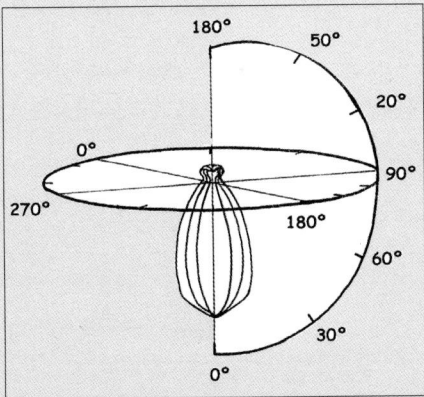

Figure 2-30. *Photometric light distribution from an industrial high-bay luminaire that directs most of its light downward with some uplight to lighten the ceiling.*

The table and graphic can tell the designer whether the lamp or luminaire is likely to be glaring, suitable for an office with computer screens, appropriate for a dark-sky-friendly parking lot, suitable for tall warehouse aisles, or an appropriate uplight for an office with a short suspension length.

For more information, see the latest edition of *The Advanced Lighting Guidelines*.[18]

3

LIGHT + VISUAL COMFORT

Light is meaningful only in relation to darkness ... It is these mingled opposites that people our life, that make it pungent, intoxicating. We only exist in terms of this conflict, in the zone where black and white clash.
— Louis Aragon

Visual Comfort

At a bare minimum, designers should create spaces in which users can perform visual tasks without discomfort, as uncomfortable lighting conditions can lead to fatigue, headaches and dissatisfaction.

Lighting conditions that are universally considered uncomfortable generally involve either an excessive range of luminances in the field of view, which results in *glare*, or instabilities of light output, which may cause *flicker*.

Visual Comfort and Glare

Glare may be discomfort glare or disability glare. Discomfort glare is glare that results in discomfort. Disability glare is brightness that reduces visual capabilities (for more information about disability glare and its impact on task visibility, see Chapter 2). Both forms of glare can be caused by light reaching the eye directly from a light source (direct glare), or indirectly from a reflective surface (reflected glare). A single bright light source or reflection can result in both discomfort and disability glare.

Points of high brightness, or *sparkle*, however, can be advantageous in some cases. For more information about sparkle, see Chapter 13.

Discomfort Glare (Including Overhead Glare): Discomfort glare is not objectively measurable, but is rather a subjective experience of discomfort. The variables involved include the luminance of the glare source, the luminance of the immediate area surrounding the glare source, the size of the glare source, and the deviation of the glare source from the line of sight.

As the luminance and size of the glare source increase, the magnitude of the discomfort increases. As the luminance of the background increases and deviation from the line of sight increases, the magnitude of the discomfort decreases, even though it may still be a source of discomfort when located above the field of view (see "Overhead Glare," Sidebar 3-2). As the number of glare sources increases, the sensation of glare also increases.

A glare source that is seen off-axis can act as a distraction. This is supported by the finding that even high-luminance sources around the fringe of and even outside the field of view can cause discomfort by producing highlights in eyebrows and on facial features surrounding the eye. This phenomenon is known as *overhead glare* and can be distracting and very uncomfortable.

As with many forms of discomfort where the reason for it is unclear, the experience of discomfort glare varies widely among individuals.

SIDEBAR 3-1. Field of View

Gaze straight ahead; there is a large visual area around the direction of the line of sight. Extend a hand to the right and move it backward until reaching the point where it is no longer visible. This is the peripheral edge of the visual field. Now repeat the process for the left, and then up and down, until all of the edges are found.

The visually observed space between these edges is the field of view.

Figure 3-1. *Illustration of field of view. (Drawing courtesy of Yukiko Yoshida.)*

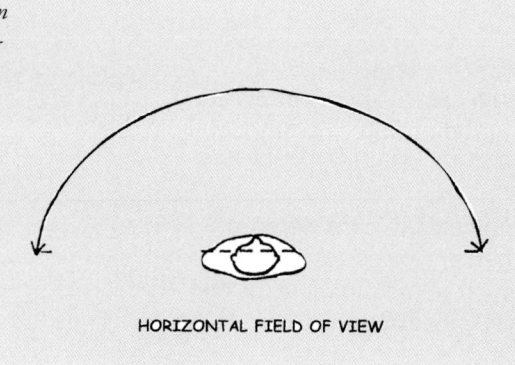
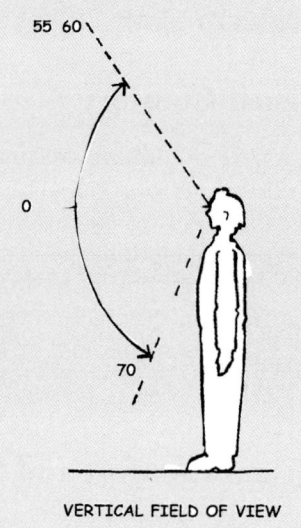

SIDEBAR 3-2. Overhead Glare[19, 20, 21]

The sensation of glare is affected by the location of the glare source in the field of view. The closer the glare source is to the direct line of sight, the greater the visual discomfort. It is commonly assumed that discomfort glare exists only if the glare source is within the field of view, or up to roughly 53° above the line of sight.[22]

By this logic, sitting in an office directly under a recessed 2x4 luminaire with parabolic louvers should not be uncomfortable because the bright fixture is technically above the field of view. However, many people are annoyed by the brightness of the lamps in the luminaire. This is called "overhead glare."

The reason: The sensation of brightness does not stop abruptly at 53°, but continues on at a continually reduced level. This sensation may be the result of direct light from the glare source scattering at the cornea of the eye and at the eyebrow. Additionally, light may be reflected into the eye from the nose, cheekbone or other part of the face adjacent to the eye.

Overhead glare can be particularly harmful in applications where users are engaged in heads-up tasks—such as typical computer applications, a classroom where the student observes the teacher and chalkboard, or a conference room where the meeting attendees are looking at each other's faces.

As a guideline for office or school lighting, bare lamps visible overhead with luminances greater than 12,000 cd/m² may be perceived as glaring. When those luminances are greater than 16,000 cd/m², most people will find them glaring. (As a point of reference, 12,000 cd/m² is a little brighter than a standard T8 fluorescent lamp; 16,000 cd/m² is a little less bright than a T5 compact fluorescent lamp.)

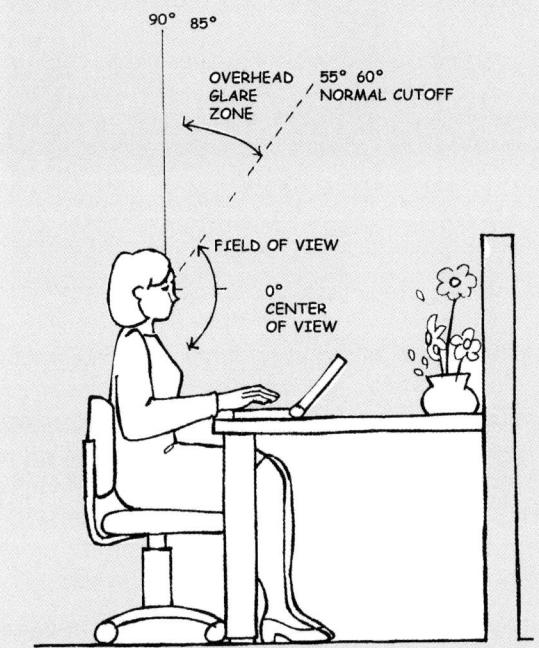

Figure 3-2. *Overhead glare zone. (Drawing courtesy of Yukiko Yoshida based on original by Naomi Miller.)*

How to Eliminate Unwanted Glare

Glare is not objectively measurable; it can be evaluated only by a series of judgments or comparisons.

The lighting industry has produced several modestly successful glare metrics, such as the Unified Glare Rating[23] and Visual Comfort Probability, which can help predict discomfort glare responses in both indoor and outdoor lighting applications.

Glare is a major concern in spaces where people spend a lot of time, as some forms of glare can result in fatigue and, in some cases, negatively impact safety and productivity in the workplace, in parking garages, or on the roadway at night.

- √ Windows are frequent sources of direct and reflected glare in workplaces, classrooms and homes. Design the room layouts and furniture arrangements to ensure users do not face a bright window for long periods of time.

- √ Faces and objects in front of bright windows may become silhouetted, making it difficult to see details in the face or object. In retail applications, objects near windows may require additional accent light to be visible. In conference rooms or classrooms, avoid locating teachers and panelists in front of bright windows.

- √ Provide blinds, overhangs, roll-down shades or other devices to mitigate periodic glaring window conditions as the sun tracks around the sky.

- √ Electric lighting near the line of sight should be low in contrast to minimize disabling glare for viewers. The luminaire should not appear very bright against the surrounding ceiling, wall or other surface.

- √ Conceal bright light sources from direct view using baffles, louvers, valances, coffers and similar media.

Figure 3-3. *Perception of discomfort glare is affected by the luminance of the glare source, the size of the source, the luminance of the surrounding area, where the source is located relative to the viewer's axis of view, and the luminance to which the viewer is adapted. (Drawing courtesy of Yukiko Yoshida based on original by Naomi Miller.)*

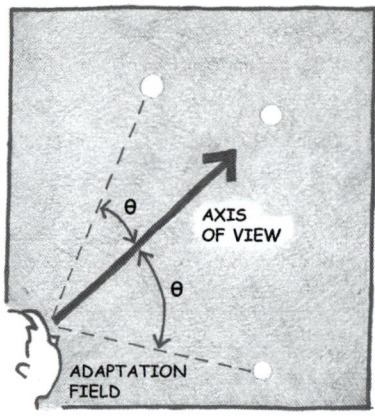

√ Luminaires that are aimed at an angle towards an object, surface or area can be particular sources of glare if the viewer walks between the luminaire and the object being illuminated. This occurs frequently in retail lighting with track lights that are aimed above a 45° angle at a merchandise display. Minimize this possibility by specifying parallel light tracks so that there are more options for the store manager to locate track heads.

√ If luminaires producing a concentrated downward light are located overhead or slightly in front of users, they may cause overhead glare, especially if the brightness of the bare lamp is visible from below.

√ If prolonged exposure is likely, consider specifying a diffusing lens to spread out the lamp's brightness—for example, a Fresnel lens in a recessed downlight, or an overlay acrylic diffuser in a parabolic-louver luminaire—or continuous dimming.

√ For office and school lighting, bare lamps visible overhead with luminances greater than 12,000 cd/m^2 may be perceived as a source of overhead glare. When those luminaires are greater than 16,000 cd/m^2, most people will find them glaring.

Figure 3-4. *Faces and objects lose detail when there is a bright window behind them. (photographer: Naomi Miller)*

Figure 3-5. *Roll-down mesh shades can reduce glare from windows without eliminating a view. Mesh shades are available in a variety of openness and colors. Darker colors are better at reducing window glare and preserving view. Light color shades brighten the room, but can also be a source of glare if illuminated by direct sunlight. (Photo courtesy of Lutron Electronics Co., Inc.)*

Figure 3-6. *(left)*
Figure 3-7. *(right)*
A corridor in a high-rise residential building for seniors, before and after lighting renovation. The left shows disabling and reflected glare from a too-bright luminaire and its reflection from the shiny floor; the right shows how a luminaire with cross-baffles minimizes luminaire brightness, redirects light to the walls, and improves visibility. (Lighting Research Center DELTA Program; Photo courtesy of © Randall Perry Photography, LLC.)

√ Overhead glare from ceiling-mounted luminaires can be mitigated by providing a high-reflectance ceiling and adding uplight. This reduces the difference between the bright lamp or lens and its surrounds.

√ For nighttime outdoor lighting, ensure that bright, visible bare lamps or bright lenses are not easily visible to drivers or pedestrians. For drivers, this often means specifying luminaires that limit the luminous intensity (candlepower) at angles of 80° to 90°, as these are the angles from which light is most often perceived as glaring from a distance. For pedestrians, it may mean specifying prismatic lensed luminaires that spread the image of the lamp's arc tube over the lens area, or reducing the wattage of the lamp. The proper solution will depend on the application.

√ Floodlights or building lights used to illuminate building façades or Automated Teller Machines (ATM) should not be tilted so that neighbors, drivers or ATM patrons find the installation glaring. With properly designed lighting, the patron and security camera will be able to have clear visibility.

√ On roadways at night, glare from oncoming vehicles, sign lighting, billboards, streetlights and building lighting can be sources of glare for both drivers and pedestrians. Factors that affect the extent of glare include luminaire light distribution and intensity, pole height, the

Figure 3-8. *Lensed luminaires flush with the bottom of the ceiling. Compare this photo with Figure 3-9, in which the same type of luminaire is mounted above the ceiling plane. (Photo courtesy of Finelite Inc.)*

Figure 3-9. *This vintage photo from the 1970s shows a "coffered ceiling system" where lensed luminaires are propped up above the plane of the ceiling to minimize the perception of glare.*

Figure 3-10. *Poorly aimed track lights can be glaring for shoppers who walk between the luminaire the illuminated display. (Drawing courtesy of Yukiko Yoshida based on original by Naomi Miller.)*

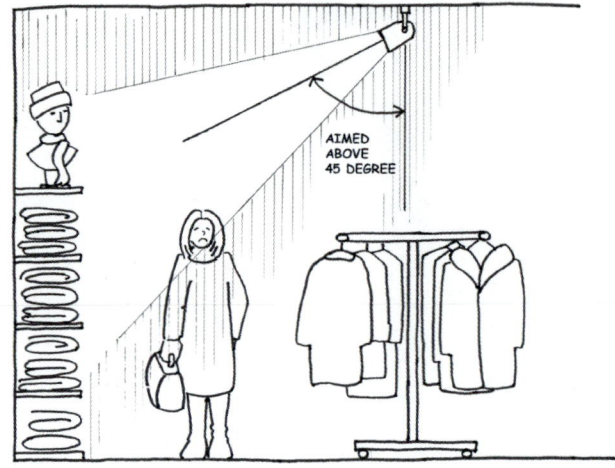

spectral content of the light, and the adaptation level of the driver or pedestrian. See IESNA RP-33, RP-20 and RP-8 for recommendations on nighttime lighting.[24, 25, 26]

Visual Comfort and Flicker/Strobe

Lamp flicker, the result of lamp instability, can be a source of discomfort in addition to glare. Individuals vary widely in sensitivity to flicker, but it is considered a cause of headaches and a trigger for migraines. Flicker can also interact with the movement of industrial machinery to produce a potentially dangerous strobe effect, in which a rotating or translating part appears to move at a different rate or appears to be not moving at all.

Flicker is the sustained and visible oscillation of a lamp's light output, usually most noticeable in peripheral vision. Many lamps that require a ballast, such as fluorescent, metal halide and high-pressure sodium, produce 120 Hz flicker (on a 60 Hz AC electrical system) in North America, or 100 Hz (on a 50 Hz AC electrical system) in most places outside North America.

How noticeable flicker is depends on the rate of flicker and the difference in amplitude of the bright/dim cycles (modulation).

Flicker should not be confused with the on/off flashing seen during the startup period of some fluorescent and high-intensity discharge (HID) lamps.

Like glare, flicker is usually a negative, unwanted condition, but can in some cases be useful and intentional. For example, strobe lights are often used to attract attention in a retail shop window or to mark an exit in a corridor during an emergency.

Figure 3-11. *(left)* **Figure 3-12.** *(right) The light-painted ceiling on the right reduces the difference in brightness between the low-bay luminaire and ceiling, making the installation appear more comfortable than in the installation on the left. (Photos courtesy of Day-Brite Lighting, Tupelo, MS.)*

Figure 3-13. *Poor (i.e., glaring) ATM lighting. (photographer: Dawn DeGrazio)*

Figure 3-14. *Better (i.e., less glaring) ATM lighting. (photographer: Dawn DeGrazio)*

How to Minimize Lamp Flicker

- √ When using fluorescent luminaires, specify electronic ballasts whenever possible; this is recommended for 60 Hz applications but is especially critical for 50 Hz applications. Electronic ballasts operate the lamps at a high frequency (20+ kHz), eliminating perception of flicker. Electronic ballasts also improve fluorescent system efficiency.

- √ In magnetically ballasted metal halide and high-pressure sodium luminaires, circuit adjacent luminaires on alternate legs of a 3-phase electrical distribution system. As long as there is an overlap of light patterns, the "off" time of one luminaire will then be offset to some extent by the light emission of the adjacent luminaire, effectively reducing the modulation of the flicker. Also consider specifying electronic HID ballasts.

- √ Avoid using incandescent lamps with diodes. Some early halogen reflector lamps relied on diodes to effectively reduce the voltage to the filament, allowing a more compact filament and better optical control. The resulting flicker can be irritating and distracting to shoppers and homeowners. Most of these lamps have been redesigned to eliminate the diode.

- √ Luminaires with higher-quality lamp sockets and effective grounding will reduce the chance of flicker caused by lamp instability.

- √ Avoid LED lighting systems that use power supplies that cause the lamp to flicker.

Figure 3-15. *It is important to consider flicker and strobe effects in factories with moving machinery. (Photo courtesy of Holophane.)*

Figure 4-1. *The eye is drawn to the sculpture because of the intriguing modeling of light. Lighting designer Janet Lennox Moyer: (photography George Gruel © George Gruel).*

LIGHT + MODELING

I never saw an ugly thing in my life: For let the form of an object be what it may—light, shade and perspective will always make it beautiful.
—John Constable

Modeling of Faces and Objects

The interplay of light and shadow can add texture and depth to a visual scene, enrich or detract from the aesthetics of surfaces, affect facial recognition and social interaction, influence perception of a space, and either improve or impair the performance of visual tasks. Light and shadow are the basic tools used for modeling of faces and objects. These tools are controllable to an extent by the designer.

Shadows can also affect task visibility; for more information, see Chapter 2.

This chapter describes how light and shadow relate to aesthetics and object and facial modeling, and recommends how to effectively use shadows when they are desirable and reduce and/or soften them when they are not desirable.

Modeling of Objects

The ability of shadows to reveal detail can be used for aesthetic as well as functional purposes. For example, shadows enhance the appearance of textured surfaces. Brick walls are visually striking when lighted from a grazing angle, as the shadow under each brick makes it appear to stand out from the plane of the wall.

Conversely, the same grazing light can reveal ugly pits and bumps in the construction of a gypsum wallboard wall. If these features should not be visually revealed, the light source can "wash" the wall from further away, eliminating the texturing effect of grazing light shadows and presenting a flat, uniform surface appearance.

In addition to revealing forms, shadows can be used to create a desirable pattern on a surface and to bring out highlights and visual interest. For example, projecting light from an uplight through the leaves and trunk of a potted plant can create dramatic jungle-like leafy patterns on the ceiling.

Figure 4-2. *Because people can walk underneath and around this monumental Bourgeois Spider sculpture in Besthoff Sculpture Garden, New Orleans, it is lighted entirely from below using ceramic metal halide in-ground luminaires. The uplight establishes two levels of lighting, producing a unique canopy effect—the overall result is playful and dramatic. (Photographer: © Richard Sexton, New Orleans)*

Figure 4-3. *Grazing light brings out the texture of this brick wall.*

Figure 4-4. *A focused halogen uplight located at the base of the blooming plant projects whimsical light patterns on the ceiling. (lighting: Luminae Souter Associates, LLC, Photography: David Winfield Wilson)*

Modeling of Faces

In addition to object and surface modeling, the relationship between light and shadow is critical for facial modeling. The pattern of light and dark on a face can make it appear friendly, attractive, young, bland, old, ugly or even evil.

Strong downlighting on a face, for example, causes unflattering deep shadowing from facial features such as eyebrows, noses and wrinkles. Overly diffuse lighting makes faces appear flat and uninteresting. In between these two extremes there is a range of flattering lighting conditions.[27]

Figure 4-5. *Facial modeling of a statue of a woman under diffused lighting. (photographer: Naomi Miller)*

Figure 4-6. *Facial modeling under diffuse lighting with sidelight. (photographer: Naomi Miller)*

Figure 4-7. *Facial modeling under downlighting only. (photographer: Naomi Miller)*

Figure 4-8. *Faces are almost imperceptible in a campus pathway lighted with low-height walkway luminaires (bollards). (photographer: Naomi Miller)*

Figure 4-9. *In another area of the same campus, post-top luminaires provide more vertical illumination on faces, making them readable from a greater distance. (photographer: Naomi Miller)*

Facial modeling, using light and shadow, can play a significant role in interpersonal communication. Good facial modeling aids visibility of eye and lip movement and gestures, which can aid understanding. This becomes even more significant in situations where the speaker is at a distance, such as in a classroom or house of worship.

Facial modeling is critical for recognition, as in the case of identification of individuals in courtrooms, parking garages, at security desks, on campus pathways and in similar situations. Facial modeling is also critical for presentation for television, videoconferencing, webcams and similar media applications; in such applications, good facial modeling may trump visual comfort for the on-camera person.

SIDEBAR 4-1. Point Source, Linear Source, Area Source

The shape and size of a light source or luminaire can affect the appearance of objects and surfaces in a space, and as such are tools for designers wishing to control the interplay of light and shadow.

Point sources are small light sources that often feature a clear outer glass bulb so that the bare incandescent filament or arc tube is visible. These sources produce dramatic highlights and sharp shadows. They are often used for retail displays and sculpture, but are undesirable for applications such as lavatory mirrors because shadows cast by eyebrows create "raccoon eyes" and accentuate facial flaws.

Many recessed luminaires with highly polished aluminum reflectors can produce beam patterns that are so concentrated that they act like point sources, even if they use larger compact fluorescent, frosted incandescent or phosphor-coated HID lamps.

Linear sources, such as straight fluorescent lamps, produce very soft shadows, and are generally desirable in workspaces where hand or body shadows may interfere with task visibility.

Area sources are large surfaces producing diffuse light, such as indirect lighted ceilings, luminous ceilings or large luminous bowl pendants. These sources almost eliminate shadows. They are suitable for uniform workspace lighting, but can produce a bland appearance when there is no other source of light such as a window, sconce or task luminaire.

Figure 4-10. *Geometric objects lighted by a point source. (photographer: © John Sutton, San Quentin, CA)*

Figure 4-11. *Geometric objects lighted by a linear source. (photographer: © John Sutton, San Quentin, CA)*

Figure 4-12. *Geometric objects lighted by an area source. (photographer: © John Sutton, San Quentin, CA)*

CHAPTER 4: LIGHT + MODELING

Figure 4-13. *Luminaires mounted close to the face of this stone fireplace "graze" the surface with light and visually amplify the stone texture. (lighting: Janet Lennox Moyer Design)*

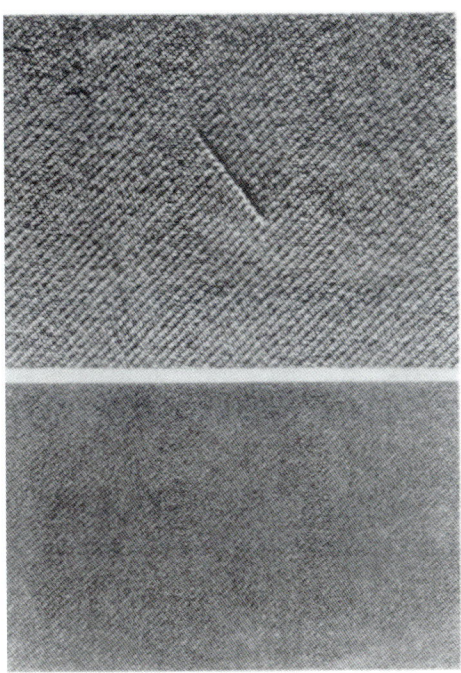

Figure 4-14. *Directional lighting (top) reveals a pulled thread unseen by diffuse lighting (bottom).*

How to Model Using Light and Shadow

√ To accentuate a strongly textured wall, mount the light sources close to the wall and aim the light almost parallel to the wall, so that the light "grazes" its surface.

√ In an industrial task where ridges, grooves or sculptural effects need to be examined, supplement the general lighting with adjustable directional task lights.

√ In museum and other applications with sculpture, shadow patterns are critical for appreciation. Most

Figure 4-15. *Sculptor Harriet Moore wanted to depict the suffering creatures in Dante's Inferno. Lighting designer Fran Kellogg Smith used strong, directional light to create shadows in the etched lines; the long shadows become as expressive as sharp black lines in an oil painting. (photographer and lighting designer: Fran Kellogg Smith)*

Figure 4-16. *Recessed 2x4 fluorescent parabolic luminaires can produce poor facial modeling if there is no other diffuse light in the space. (Photo courtesy of Litecontol.)*

Figure 4-17. *Good lavatory lighting provides diffuse lighting with sufficient vertical illuminance in this demonstration vignette. (Photo courtesy of Energy and Technology Center, Sacramento Municipal Utility District.)*

museum spaces should have the ability to aim focused point sources at artwork and bring out details.

√ For theatrical and photographic lighting, there are critical angles and combinations of lights for facial modeling that maximize visibility and improve appearance.[28] For most workplaces, one needs only to avoid poor conditions.

√ Pronounced shadows make faces appear unpleasant or unnatural, while soft shadows can make faces appear more pleasing and identifiable. Avoid using strongly directional light in places where the appearance of faces is important. Examples include recessed downlights with a concentrated distribution and fluorescent luminaires with highly specular (shiny) parabolic louvers. If directional luminaires will be used, consider luminaires with lenses, diffusers, louvers with a semi-specular finish, or white reflecting surfaces to diffuse distribution; shadowing can also be reduced or softened by adding a diffuse layer of light by washing walls, uplighting ceilings or applying luminous sconces to walls.

√ Consider luminaires with a wide beam spread and mount luminaires closer together to overlap lighting distributions at head height.

LIGHT + COLOR

"Yes," I answered you last night. "No," this morning, sir. I say: Colors seen by candle-light will not look the same by day. – Elizabeth Barrett Browning

Color Appearance (of Objects, People or Light Sources)

Color is a critical element that has an integral relationship with light: The spectrum of the light source affects how humans will perceive the colors they see in the lighted environment. Color is one of the most definitive examples of where simple lighting choices can have dramatically different effects on how a space is visually experienced by its users.

Color quality encompasses many issues, including:

- The color of the light source—whether it's "warm" or "cool" in appearance;
- Color rendering of room finishes and objects—whether the colors are attractive and rendered accurately;
- Color appearance of skin tones—whether faces look ruddy, jaundiced, pale, etc. under the light; and
- Color contrast—describing the visibility of the task detail against its background.

The designer's lighting-related color choices can affect how healthy people appear in a space, task performance, communication, perception of merchandise and artwork, and overall presentation of a space's interior finishes and furnishings. The color characteristics of a light source can affect perceived brightness. For example, a light source deficient in red wavelengths may deliver high illuminances on a red apple, but the apple will still appear dull or "grayed out."

Two metrics are used to describe the color quality of a light source used in architectural applications: *correlated color temperature* and *color rendering index*. These metrics allow lighting designers to compare the color quality of light sources and both control and predict the color impact of different sources.

The designer must first determine how important color is in the application. Some light sources offer medium or poor color qualities but higher *efficacies* (lumens of light output per watt of electrical input). Other light sources offer superior color qualities but may present a higher installed cost. The designer must determine the application need, then identify and com-

Figure 5-1. *Niches with downlights lamped with 2700K, 3000K, 3500K and 4100K lamps. (Photo courtesy of Energy and Technology Center, Sacramento Municipal Utility District.)*

pare light sources—assessing potential tradeoffs in appearance, performance, efficacy and cost—to select the most appropriate source.

This chapter describes the impact that lighting choices have on color perception, the metrics used to express color quality, and how to design to achieve good color quality in a space.

Color Temperature

The color temperature of a light source, expressed in kelvins (K), indicates the color appearance of the light source itself and the light it emits. Light sources are generally classified as "cool" (approximately 4000K or greater), which appear bluish white; "neutral" (approximately 3500K), which appear white; or "warm" (approximately 3000K or less), which appear yellowish white.

Warm light sources are more heavily laden with red and orange wavelengths, bringing out some flesh tones and richer appearance in objects that have warmer colors. Cool light sources are more heavily laden with blue and green wavelengths, enriching the visible color appearance of blue and green objects.

Whether the light is warm or cool in appearance, the light seems to have a "natural" and acceptable appearance if its color coordinates lie along the "black body locus" of the CIE Chromaticity Diagram (see Chapter 4 in the 9th Edition of the IESNA *Lighting Handbook* for additional technical information.) This means that unless a dramatic colored effect is intended, normal "white" electric lighting should not appear too purple, too green, too blue, too orange, etc.

Otherwise, no firm guideline is offered, as color temperature selection is driven by psychological concerns that are primarily cultural and possibly climate-related. For example, many Americans associate warm light sources with comfort and relaxation because they grew up with incandescent lamps in their homes, while many people in Mexico and Asian countries prefer very cool light sources. As another example, people in northern latitudes may prefer warmer light sources, while people in southern latitudes prefer cooler light sources, possibly as a psychological compensation for the predominant weather in these latitudes.

SIDEBAR 5-1. Color Temperature

Lighting designers often refer to the color appearance of a light source as being either "warm" or "cool," similarly to the way interior designers describe palettes of paint or fabric. "Warm" refers to light with a yellowish or peach-colored cast, while "cool" refers to light with a bluish cast. These qualities are expressed using a metric called the color temperature of a light source. The lower the color temperature, the warmer the light source appears; the higher the color temperature, the cooler the light source appears.

The color temperature of a light source is expressed in kelvins (K)—just like degrees Celsius, but beginning at absolute zero. To determine color temperature, a theoretical reference source called a blackbody radiator is used. Imagine a block of iron which is completely black at room temperature, but as electricity is passed through it, it heats until it glows reddish-orange, similar to a heating element on top of an electric stove. As the temperature increases, it becomes orange, then yellow and so on until it glows bluish-white. At any point during this heating, the block's temperature could be measured and its color appearance at that time recorded. Above 5000K, the standards community uses agreed-upon forms of daylight to set these reference color temperatures.

The color produced by a lamp is compared with this reference source to determine its color temperature. If it is an incandescent lamp, which uses a heated element similar to a blackbody radiator, the temperature of the reference source that matches the lamp's color becomes the lamp's "color temperature." If the lamp is a fluorescent or HID lamp, which does not use a heated element, the value is approximated through correlation, and is therefore called "correlated color temperature" (CCT). Typically, in the field, color temperature and correlated color temperature are both referred to as "color temperature."

Approximate color temperature values for sample light sources are shown below.

10,000K –	blue-white light, like north daylight
6500K –	very cool, almost bluish white light, like daylight from sun and sky at noon
5300K –	noon sunlight
4000K –	cool shade of white light, like the cooler fluorescent or most metal halide lamps
3500K –	neutral shade of white light
3000K –	warm shade of white light, like halogen incandescent lamps, or some warmer fluorescent lamps and some metal halide lamps
2700K –	yellow-white light, like standard incandescent lamps
2200K –	orange-yellowish light, like standard high pressure sodium
1800K –	orange-ish, like candlelight, or light from the setting sun
1200K –	orange-red
800K –	dull red

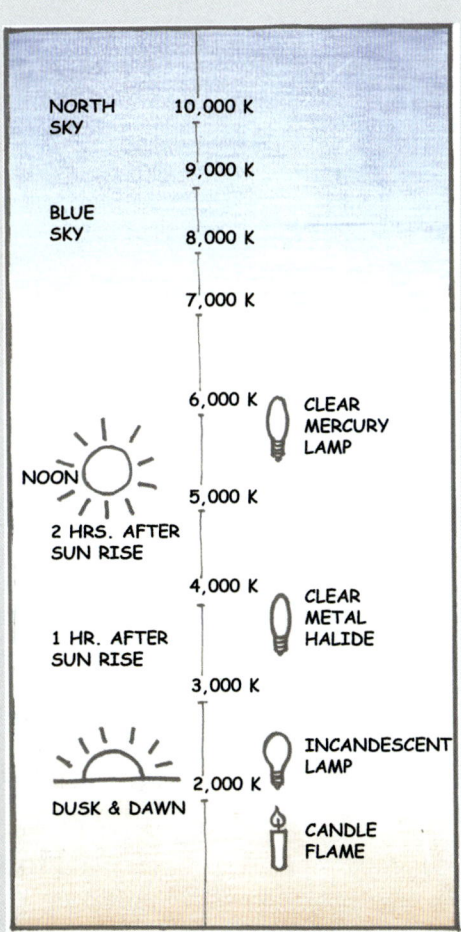

Figure 5-2. *Correlated color temperature scale. (Drawing courtesy of Yukiko Yoshida.)*

Figure 5-3. *Marketplace apples as rendered under high and low CRI conditions. (Photo courtesy of General Electric Company.)*

Color Rendering

Color rendering is the ability of a light source to make skin tones, plants, room finishes, or fabrics look "natural." It is a function of the spectral power distribution of the light source and the color reflectance properties of the room finishes, object or face.

The color rendering ability of light sources is quantified using the color rendering index (CRI, or R_a outside of North America), a mathematically derived metric that uses a maximum 100 scale. Basically, the closer to 100, the better the color rendering.

A critical step to comparing the CRI ratings of different light sources is to ensure that they have the same color temperature. Otherwise, the comparison is meaningless.

CRI is an imperfect metric for several reasons,[29, 30] but it is popularly used and can be an effective predictor when the designer cannot test the actual object with the actual lamp. To gain a true sense of a light source's color rendering ability, however, there is no substitute for seeing firsthand how a given light source illuminates different objects. For example, asparagus viewed under incandescent light can be disappointingly dull, while red apples may appear very appetizing. Navy blue drapes in a lobby may appear gray under incandescent lighting, but livelier under fluorescent or metal halide lamps.

How to Achieve Good Color Quality

√ Select room finishes, colors and fabrics for a project by first examining them under the light of the same lamps being proposed for the project, as both color

SIDEBAR 5-2. Color Rendering Index

Because the color appearance of the lamp itself does not reveal how well it renders colors, the color rendering index (CRI) is used to help quantify a lamp's ability to produce color in objects.

Lamps are assigned a CRI rating based on how similarly they render a set of eight object colors compared to a reference light source of the same CCT. For lamps with a CCT below 5000K, the reference source is a blackbody radiator (similar to a tungsten filament heated to different temperatures and giving off different colors of light, depending on the temperature). For lamps with a CCT greater than 5000K, the reference source is a mathematical model of daylight derived from measurements of the daylight spectrum.

CRI is measured on a scale of 0-100. Lamps that render the eight test colors very similarly to the test source will have a high CRI. Conversely, lamps with a low CRI produce a large color shift when compared to the standard.

Because the reference for CRI changes with color temperature, it is only valid to compare the CRI ratings of different lamps if they have similar color temperature values.

Lamps with a higher CRI are *generally* better at making objects and surface colors appear more colorful and natural than lamps with a lower CRI. However, there is no substitute for evaluating color with one's own eyes. Because of the way this measure is computed, a lamp that is visually preferable does not always have the highest CRI value.

Figure 5-4. *The eight object test colors for calculation of CRI.*

temperature and CRI—and even the reflections from surrounding colored surfaces—will affect the color appearance in the lighted space.

√ Properly lighting skin tones is a critical color issue for many applications.

- A given light source may render the colors of objects and surfaces in the space spectacularly, but make people appear unnatural. Some light sources can even make skin tones appear gray.

- Where pleasant skin tones are desirable, a high CRI is not sufficient in choosing the light source. One research study suggests that a neutral color temperature is also important. Study participants from a range of racial groups found that they liked their skin appearance best when the color temperature of the light was between 3000K and 4100K, with most choosing a neutral color temperature of 3500K. Very warm (2850K) and very cool (5000K) light sources received the least favorable responses from participants.[31]

- When testing light sources, the designer's own hand can be an excellent reference for judging color rendering of skin tones, especially if the designer has become familiar with his or her own hand appearance under different light sources.

√ Be aware that color contrast will be affected by lamp color choices. For example, the color contrast of letters against a background may make a sign more visible under one light source than another. In a warehouse application, it would be more difficult to read, for example, orange lettering on a brown cardboard box under high-pressure sodium lamps than metal halide lamps.

√ Certain colors can be surprisingly distorted by light sources. For example, some purple glass appears pale blue under fluorescent lighting because fluorescent is deficient in the far red wavelengths that give purple its special richness. Halogen light or daylight bring out that purple. On the other hand, incandescent light is poor in blue wavelengths, so it may be difficult to distinguish blue from black socks under low levels of incandescent light.

√ When mixing two or more different light source types in the same space, a unified color of light within the space is usually desirable. In other words, the color temperatures and color rendering abilities of the light sources should match as closely as possible unless the designer wants to accentuate differences through color contrast.

- How warm or cool a lamp appears is intensified when viewed next to a dissimilar light source. The eye quickly picks out a single warm-white fluorescent lamp installed in a room with cool white lamps. Similarly, a wall sconce with warm 2700K compact fluorescent lamps can appear unnaturally pink in a room lighted with 3500K (neutral white) or daylight (very cool) general lighting.

- Where compact fluorescent or T5 lamps are used, consider companion T8 fluorescent lamps with a matching color temperature and a minimum of 80 CRI. Compact fluorescent lamps and T5 are only available with a minimum of 80 CRI and the price difference between a 70 CRI and an 80 CRI T8 lamp is minimal.

- In some cases, the designer may wish to deliberately use a lamp that doesn't match the ambient color so as to achieve a desired color effect. In this event, ensure the effect is significant enough that it does not simply appear to be a mistake. Addi-

Figure 5-5. *Consider the color effects of non-electric light sources that may be present, such as daylight. (Photo courtesy of Selux.)*

tionally, the maintenance staff responsible for lamp replacement should be notified of the lamp schedule and its rationale so that the effect can be preserved after initial installation.

√ When considering color balancing, keep in mind other non-electric sources of light in the space such as daylight, candles and hearth fires.

√ Some designers find it useful to examine the spectral power distribution (SPD) of the light sources they use. SPD shows the visible light spectrum and the wavelength composition for the light from the lamp. Some sources provide a continuous power distribution. Other light sources have spiked or discontinuous power distributions; the spikes indicate that the light is stronger in revealing certain colors. If a particular SPD is sparse—a few narrow lines—then predicting the color rendering ability of the light source will be difficult; in this case, it is advisable to use a live demonstration to determine the color appearance of people and objects. If the SPD is broad—emitting energy continuously across the spectrum—and in particular the red portion of the spectrum is substantial, then a live evaluation is considered less critical.

√ Select a CRI rating >50 if occupants need to roughly identify or differentiate colors in the space. This is applicable in many industrial applications, and frequently applicable for outdoor spaces, such as parking lots, where visitors need to identify their vehicles by color.

Figure 5-6. *Normal warehouse and assembly tasks call for >50 CRI. (Photo courtesy of Acuity Brands Lighting.)*

√ Select a CRI rating >80 in spaces where people communicate with each other regularly, or if food is involved; this includes offices, conference rooms, schools, restaurants and cafeterias.

Figure 5-7. *Office space tasks generally call for >80 CRI. Epic Data Office, Richmond, British Columbia. (lighting: Ledalite Architectural Products)*

√ Select a CRI rating >80 for applications where pleasant skin tones are important or where appreciation of artwork or room finishes are primary issues; this includes conference rooms, interview spaces, building lobbies, museums and galleries and homes.

√ Select a CRI rating >80 for retail applications where color is critical to the shopper, such as clothing stores, supermarkets and furniture stores.

Figure 5-8. *Foods look much more appetizing at >80 CRI. Market of Choice. (Photo courtesy of Con-Tech Lighting.)*

√ Select a CRI rating >90 if color-matching tasks are being performed, such as mixing paints or matching fabric dye lots. Incandescent/halogen lamps are often not acceptable in these applications because, despite high CRI ratings, they are relatively deficient in the violet and blue parts of the color spectrum.

Figure 5-9. *If color is critical for matching of dye lots, CRI should range between 90 and 100. Daylight can also be a good choice.*

PART TWO: LIGHT + ECONOMICS & ENVIRONMENT

God is in the details.

– Ludwig Mies van der Rohe

LIGHT + COST

Lack of money is no obstacle. Lack of an idea is an obstacle. – Ken Hakuta

Cost, Both Initial and Maintained

Lighting design projects are often constrained by the designated budget. The designer, to an extent, has control over both the initial and maintained costs of the lighting system.

Initial cost includes all costs associated with producing the initial lighting installation. This can include all design fees, equipment purchasing costs, installation labor costs, etc.

Maintained cost includes all costs associated with operating the lighting system over time. This can include all lamp and ballast replacement costs, maintenance labor costs, lamp and ballast disposal costs, energy costs, etc.

In many cases, it can be desirable to compare the *life-cycle costs* of different lighting systems to determine best value. The life-cycle cost for a given lighting system includes both initial and maintained costs over the projected life of the system, resulting in a total cost of ownership for the system.

This chapter provides considerations for maximizing value.

How to Maximize Value on a Given Budget

√   Know the project's budget prior to beginning the design. The alternative may be to design the project twice—once for a fee, and once for experience.

√   Lighting budgets are often set with an unrealistic goal. An experienced designer can often examine the budget, evaluate whether the allotted sum can cover the lighting system expected by the client, and negotiate for a higher budget if necessary. This is better done earlier than later in the project.

√   Smaller budgets and good lighting solutions are not mutually exclusive. Creativity and coordination with the design team can make it possible to achieve good results with simple, inexpensive equipment. Fluorescent striplights, inexpensive track lighting and even metal halide wallpacks can be concealed behind beams or fascias, or covered with a metal shroud to utilize their functionality but mask their utilitarian appearance.

Figure 6-1. *This family room uses a simple and inexpensive fluorescent striplight mounted behind the wood beam. (© 1994 Kathryn M. Conway)*

Figure 6-2. *This office space uses economical fluorescent industrial luminaires inverted and mounted as uplights. (lighting: Patrick B. Quigley & Associates; photographer: Toshi Yoshimi)*

√ Occasionally, the budget has been set so low that compliance will result in a lighting system that cannot satisfy client needs or design goals. If the designer is aware of this, it is his or her responsibility to report this to the client.

√ Specify energy-effective lighting equipment and controls. Available lighting technology can provide aesthetics, versatility and performance while minimizing operating costs. This requires the client to think beyond initial cost and adopt life-cycle costing—financial decisions based on the total cost of ownership of the lighting equipment rather than simply the initial cost to purchase and install.

LIGHT + MAINTENANCE

Hard work spotlights the character of people: Some turn up their sleeves, some turn up their noses, and some don't turn up at all. – Sam Ewing

Maintenance and Change

Although a lighting design may be ideal the day it is installed, the performance of all lighting systems deteriorates over time. Luminaire output decreases as lamps age, for example, and both luminaire and reflective room surfaces collect dirt and dust. Environmental conditions may change in the application. In landscape designs, for example, plants will grow, and the landscape may be altered by design; the use of the space may also change. In addition, automatic lighting controls may need to be recalibrated to continue being responsive in the face of changes in the space. For these reasons, all lighting systems must be maintained.

This chapter provides recommendations for lighting maintenance, including planned lighting maintenance. For more information, see IESNA RP-36, *Recommended Practice for Planned Indoor Lighting Maintenance*.

How to Design for Maintenance

√   When designing, consider how the system will be maintained. For example, if the designer would find it risky to change the lamp in a recessed incandescent downlight 10 meters (30 feet) above an escalator, then it may not make sense to place this luminaire in that location. Difficulties in accessing some luminaires may also affect lamp and ballast choices—generally, it is advisable to select components that provide long service life. For some hard-to-reach luminaires, the designer may need to discuss with the client maintenance resource needs, such as lifts, ensuring doors are large enough to bring a lift into a room, or adding catwalks for access.

√   Plan for predictable loss in light output and illuminances over time due to luminaire dirt depreciation (LDD) and lamp lumen depreciation (LLD). This generally entails designing the lighting system to provide higher illuminances when it is new, so that over time,

Figure 7-1. *Scheduled luminaire cleaning can improve maintained illuminances. (Photo courtesy of Colorado Lighting.)*

Figure 7-2. *Scheduled group relamping can economize on lamp replacement labor. (Photo courtesy of Colorado Lighting.)*

as lamp and luminaire output deteriorates, minimum design illuminances will be still be delivered.

√ Recommend planned, periodic luminaire cleaning to improve maintained illuminances. If less dirt depreciation is expected, the lighting system can be designed for light levels that are closer to the target light levels, which can reduce capital and operating costs.

√ If the economics make sense for the application, recommend group relamping, which is the practice of replacing all lamps in a population at the point when their failure rate accelerates (usually at 70-80% of rated service life in the case of fluorescent lamps), instead of individually as they fail, to increase the efficiency of labor utilization. It is often more economical to replace all lamps, even lamps that are still operating, as a group on a planned basis rather than individually as they fail.

√ Select as many luminaires as possible that use the same lamp type. If fewer lamps are used in the project, it can simplify lamp replacement and reduce storage costs. The easier the lighting system is to maintain, the more likely it will that the maintenance staff will maintain it properly and sustain the original design vision.

√ When lamps fail, they are often replaced by maintenance staff that use the first, and usually the cheapest, lamp they can find that fits the socket. Make the maintenance department aware of any special lamps used, particularly when other lamps will fit the same socket. If high-lumen 32W T8 lamps are specified, for example, ensure that the maintenance department does not replace them with standard 32W T8 lamps, as this will change illuminances in the space and af-

CHAPTER 7: LIGHT + MAINTENANCE

fect the design. Recommend that additional lamps be purchased at the time of installation so that the first failed lamps are sure to be replaced with the same lamps.

√ Consider specifying adjustable accent luminaires that can be re-aimed in response to changing points of interest or changing needs

√ Design additional flexibility into the lighting system if it is anticipated that there will be changes to how the space is used. Lighting installations can experience significant changes over time. Landscape lighting designers, for example, understand that as shrubbery grows, the lighting must able to accommodate the change in size. In interiors, lighting systems may grow inefficient or obsolete, furnishings and appliances get moved, fashions change and area functions grow, shift or are replaced altogether. Change is normal (particularly in retail) and in some spaces, flexibility can be designed into the system to accommodate future needs.

√ Planned maintenance operations can be economically combined with energy-efficient lighting upgrade operations.

√ Recommend consultation to reconsider the lighting design in the event that a space is reconfigured. Office furniture sizes, locations, surface finishes and other characteristics can all have a significant effect on illuminances and light patterns in a space.

√ Emergency lighting systems must be tested periodically to comply with local codes.

√ Commission lighting control systems after installation to ensure proper operation, and establish a schedule to recalibrate controls and re-commission the system over time (see Chapter 10).

√ The owner of the lighting system must comply with all federal, state and local regulations regarding disposal of spent lighting components. This may include fluorescent or HID lamps, ballasts or emergency battery packs or exit signs.

LIGHT + ENERGY

Faith is like electricity. You can't see it, but you can see the light. – Unknown

Energy Use

According to the International Energy Agency, electric lighting consumes 19% of total global electricity production.[32] The Lawrence Berkeley National Laboratory has estimated the global energy operating cost of lighting to be approximately $230 billion per year, of which about $70-$95 billion could be saved using today's energy-efficient lighting technologies.[33]

Responsible lighting design is energy-effective. Energy-effective design has two dimensions. First, all other things being equal, it provides a high level of energy efficiency, which contributes to a lower lighting system ownership cost. Second, it is effective—that is, it provides desired quantity and quality of illumination.

Energy-efficient lighting is often required by energy codes. However, as with all codes and standards, they provide only a minimum standard. Energy-efficient lighting can reduce carbon and other air emissions at power plants.

This chapter describes considerations for designing for energy efficiency.

Design Considerations for Minimizing Energy Use

√ *Energy-effective design.* In order to be energy *efficient* in the context of lighting design, one needs to remember to use light *effectively* for the needs of the task and the space. "Energy effective" lighting design considers many factors in developing a solution. It is not only the components of a lighting system that make the difference, it is also how these components interact in a system and how the system interacts with the space and its users.

√ *Efficacy of light sources.* Energy efficiency begins with the light source. Light sources can be evaluated using *efficacy*, or lumens of light output per watt of electrical input. Examples of efficacy ratings include 13.3-

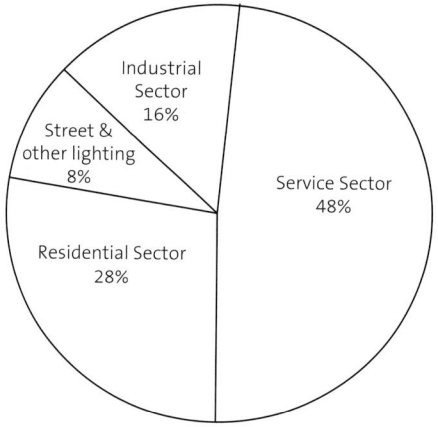

Figure 8-1. *Estimated global electricity use by end-use sector, according to the International Association for Energy-Efficient Lighting and the Lawrence Berkeley National Laboratory.*

Figure 8-2. *Even when two rooms are lighted with the same watts, one can use energy more effectively than the other. (Drawing courtesy of Yukiko Yoshida)*

20.7 lumens/W for halogen IR PAR lamps, 58.1-64.0 mean lumens/W for 13-42W compact fluorescent lamps, 87.5-92.2 mean lumens/W for 32W T8 linear fluorescent lamps, and 51.0-85.0 mean lumens/W for 50-400W pulse-start metal halide lamps.[34, 35]

√ *Efficiency and output of ballasts.* If a ballast is present in the lighting system, electronic ballasts should be selected whenever possible. Newer-generation high-efficiency electronic ballasts increase efficacy even further. Some ballasts are available with different outputs—low, medium and high ballast factors—which provide light output and energy savings flexibility.

√ *Photometric distribution of lamp.* Light distribution begins with the lamp; light is then passed through the luminaire and ultimately enters the space. Some lamps provide a high degree of control over distribution, while others do not. For example, using a 20W compact fluorescent lamp for retail accent lighting may be less energy-effective than using a 20W halogen lamp, as the halogen reflector lamp is much more effective at delivering illuminance on merchandise, resulting in fewer lamps being needed.

√ *Photometric distribution of luminaire.* When non-reflector lamps are used, the photometric distribution of the luminaire becomes more important. Below are key considerations:

- Luminaire efficiency—the percentage of light that exits the luminaire rather than being trapped inside.
- Coefficient of Utilization (CU), or the percentage of light delivered to the workplane given the specific surface reflectances and size and shape of the room.
- Glare and the potential for light falling where it is not desired or needed.
- The photometric distribution information available from luminaire manufacturers.
- The placement of the luminaire in relation to the occupants and architecture.

√ *Room reflectances.* Energy efficiency is directly related to and dependent upon surface reflectances in the space. The best selections of lamp, ballast, luminaire and layout can be neutralized if room surfaces are dark. This is where different design disciplines need to collaborate to achieve overall design success. In general, to achieve an energy-effective design, large wall surfaces, partitions and furniture should be light-colored, while accents can be light-, medium- or dark-colored.

√ *Controls.* Lighting controls can make even a high-wattage lighting system more energy-effective by using dimming or automatic switching to reduce both the power required for lighting operation and the amount of time it is in operation (energy = power x time). For more information, see Chapter 10.

Figure 9-1. *HGA Architects, Engineers and Planners recently earned a LEED-Commercial Interiors (CI) Silver rating for its new offices in Milwaukee, which incorporated daylighting, views, low energy consumption and individual task lighting for occupants. (lighting: Jill Cody; photographer: Jill Cody)*

LIGHT + ENVIRONMENT

An optimist is a person who sees a green light everywhere ...
– Albert Schweitzer

Environmental Considerations

Commercial and residential buildings in the United States currently consume 65% of domestic electricity production, 12% of domestic potable water and 40% (or three billion tons) of raw materials globally, while causing about 30% of total greenhouse gas emissions and 136 million tons of construction and demolition waste each year.

This model is not sustainable. Construction depends on an enormous amount of resources, and yet these resources are in finite supply and are steadily diminishing.

Sustainability, sometimes called *green design*, is an approach to design that seeks to minimize the environmental footprint of buildings.

In 1987, the United Nations World Commission on Environment and Development (Brundtland Commission) defined sustainability as "meeting the needs of the present without compromising the ability of future generations to meet their own needs."

For designers of lighting systems, this entails providing a quality lighted environment for people who use the space, with minimal consumption of resources. Green design, as it applies to lighting, involves reducing pollution and waste. Pollution and waste are generated during the manufacture and transportation of products, emitted into the atmosphere as byproducts of power plant electric generation, released as unutilized energy such as heat and stray light into the environment, and generated by the disposal of finished materials such as spent lamps and ballasts.

This chapter describes how lighting choices can affect sustainability.

Efficiency in Whole, Not Part

Because most of the environmental impact of lighting occurs during the usage phase of a building, energy efficiency is a critical environmental issue.

When designing for efficiency, emphasis should be placed on achieving overall lighting system efficiency rather than focusing solely on the relative efficiency of each individual component in the system. For example, an individual component may be relatively efficient, but a high density of such components can result in poor overall efficiency.

Rating Systems Provide Benchmarks

In an effort to help designers navigate the interconnected issues of sustainable design, various organizations have developed rating systems to rank how sustainable a specific project is compared to other projects. The most commonly used are the Leadership in Energy & Environmental Design (LEED®) rating systems, as developed by the U.S. Green Building Council (http://www.usgbc.org).

Lighting's Role in Sustainability

Although all sustainable design issues should be considered on a quality lighting project, there are three where lighting plays a significant role: minimizing energy consumption, minimizing light pollution, and incorporating daylight into designs.

As these topics are covered extensively in other chapters in Light + Design, the remainder of this chapter will present other steps designers can take to make their lighting designs more sustainable.

Cradle to Cradle Products

Life-cycle assessments of lighting products use the concept of embodied energy. Embodied energy reflects the total amount of energy that goes into a product from beginning of production (including all component parts, such as mining of metals and manufacture of plastics) through transportation of the component materials and finished goods, installation, useful life and on to disposal—in other words, "cradle to grave."

The philosophy of "cradle to cradle" design, however, is that all materials used in a product are recycled back into the raw materials to build new product. While currently rare, cradle-to-cradle manufacturing is becoming a reality for a growing number of pioneering architectural products.

Think Globally, Design Locally

To design for sustainability, a designer needs to think globally by responding locally. Each project site has unique local conditions with respect to solar angles, climate conditions, species impacts, available resources and regulations, to name a few. Design solutions that work well in Montreal, Quebec may not address issues in Phoenix, Arizona, and vice versa. Each design solution, therefore, needs to be customized to the individual project.

Lamps and the Environment

Lamps are of particular concern because they may contain lead and mercury. The elimination of lead solder in lamp bases is becoming common practice. However, mercury is a fundamental ingredient to producing light in fluorescent and HID lamps. Ironically, the use of mercury in lamps can contribute to other sustainable design benefits such as energy efficiency and lamp longevity. The question is how little mercury can be included in a lamp and still produce acceptable light output and lamp life. Lamp manufacturers are addressing this question and producing lamps with significantly lower levels of mercury than in previous years.

In an effort to regulate mercury in the waste stream, the United States government created the Toxicity Characteristic Leaching Procedure (TCLP) test. Lamps that fail the TCLP test are classified as hazardous waste and must be handled according to stringent regulations, whereas lamps that pass the TCLP test may be allowed to be processed as solid waste, subject to state and local regulations (a number of states are more restrictive—e.g., do not allow any fluorescent lamps in solid waste landfills). The most "green" solution is to recycle mercury-containing lamps, even those that pass the TCLP standard.[36] A list of lamp recycling companies can be found at www.lamprecycle.org.

How to Support Sustainable Design Goals with Lighting Choices

- √ Energy efficiency is by far the greatest contribution that lighting can make to sustainable design. Designs should be the lowest reasonable connected load and save energy over time via the use of lighting controls.

- √ Consider all LEED options related to lighting.

- √ Encourage manufacturers to develop products that can be easily recycled back into use in manufacturing new products.

- √ Specify materials and products that are durable and long lasting.

- √ Customize design solutions to local conditions such as solar angles, climate, species impacts, available resources, regulations and other relevant concerns.

- √ Specify lamps with a long service life.

- √ Specify fluorescent and HID lamps with low mercury content.

- √ Encourage the owner to recycle spent lamps and ballasts.

LIGHT + CONTROL

Put out the light. — last words of Theodore Roosevelt

Turn up the lights. I don't want to go home in the dark.
— last words of O. Henry

Controls for Energy, Flexibility and People

The use of advanced lighting controls can provide a higher level of energy effectiveness, support flexibility of use of a space, and increase occupant satisfaction. A good controls design can help deliver the right amount of light where it is needed and when it is needed.

This chapter examines how to increase flexibility and energy effectiveness, while potentially influencing occupant satisfaction, with lighting controls.

Controls for Energy

Controls can support energy management goals by increasing the energy effectiveness of the lighting system. According to the New Buildings Institute, lighting controls can reduce energy consumption by as much as 50% in existing buildings and by at least 35% in new construction[37]:

- Lower operating costs through reducing lighting system power requirements and operating hours
- Lower operating costs through reducing lighting system power use during peak electrical demand hours

Energy consumption can be reduced either through human initiative or automatic operation that in turn involves a change to the lighting condition based on either a time-based or threshold-based event.

An example of a time-based event is scheduling, in which the lighting system is shut off automatically at a preset time of day.

Examples of threshold-based events include daylight harvesting, in which the lights are switched or dimmed due to a sufficient increase in illuminance in the space from daylight[38], and occupancy sensing, in which the lights are switched based on the detected absence or presence of people in the space.

Figure 10-1. *(top)*
Figure 10-2. *(bottom)*
An architectural control system, combined with a multi-layered lighting system, allows the staff to customize the mood for any event scheduled in this ballroom. (Photos courtesy of Lutron Electronics Co., Inc.)

Controls for Flexibility

Controls can support flexibility of use of a space by enabling:

- Adaptation of lighting for multiple uses of a space
- Adaptation of lighting to evolving space needs

Hotel ballrooms are used for corporate holiday parties, starving artist shows, college entrance exams, real estate seminars, Bar Mitzvahs, professional society conferences, romantic wedding receptions, car shows and the occasional ball. Can one lighting system satisfy all these needs? Probably not.

One way to satisfy these diverse needs is to design the lighting with different layers of light. In this case, the ballroom could have three or more lighting layers: high-illuminance general lighting using fluorescent lamps, recessed incandescent downlights for lighting table settings only, and decorative chandeliers and wall sconces.

By switching or dimming all of these layers separately, the hotel can create a romantic, festive atmosphere for a wedding party; a bright, businesslike space for seminars; and a dark space with note-taking light for conference data projector presentations. A lighting control system with dimming and multiple-zone switching makes this flexibility of use possible.

There are other reasons for achieving flexibility with lighting controls. A retail store may want to have a different light level for day and night operation, since it takes less light at night for a store to look open and attractive than when its customers are adapted to the bright outdoors. Decorative

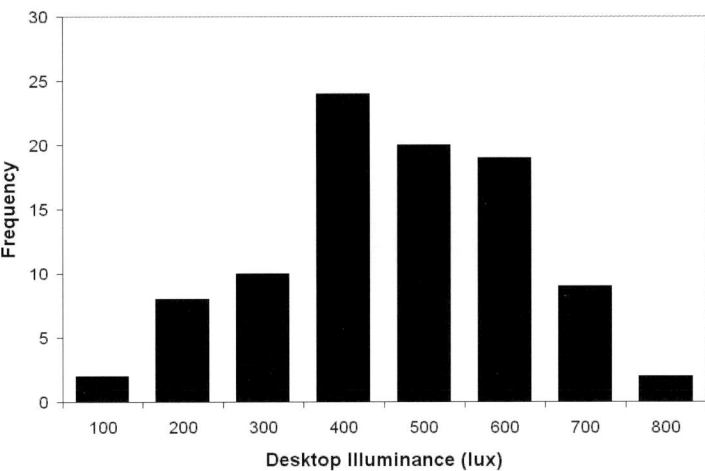

Figure 10-3. *In a realistic office laboratory, 94 research participants were instructed to use dimmers to control various luminaires associated with their workstation. This histogram illustrates the desktop illuminances resulting from their choices, revealing a wide range of individual preferences. (illustration created by Dr. Guy Newsham; copyright National Research Council of Canada)*

landscape lighting may be switched separately from the parking lot lighting so that it can be shut off at an earlier time of night.

Putting different lights on separate control circuits increases flexibility. Adding dimming to one or more control circuits increases the range of lighting combinations available. Dimming can save energy and prolong halogen and incandescent lamp life. Assigning groups of luminaires to different control zones and dimming them at different rates can reduce the number of different lamp types used on a project.

Controls for Occupant Satisfaction

Controls may support occupant satisfaction of use of a space by enabling:

- Mood setting
- Individual control of local illuminance and brightness on vertical surfaces

Mood setting is primarily accomplished via dimming, which enables adjustment of light levels to achieve a desired ambiance.

Research suggests that users of lighting systems have a wide range of preferences of illuminances.[39] Providing user control of illuminances can increase satisfaction, mood and comfort of office workers, according to research.[40, 41, 42, 43]

For example, an office workstation may have task lights for the user to switch on whenever he or she needs additional work light; or the workstation may enable the user to dim light from an overhead fixture using a handheld remote or interface on his or her personal computer.

How to Incorporate Controls into a Project

√ Should the load be switched or dimmed?

- Switching—turning the lights on or off—is primarily used for energy management purposes. Strategies include occupancy sensors, daylight harvesting control, bi- and multi-level switching, and scheduling. Automatic lighting shutoff for certain buildings is now required by most energy codes in the United States. Switching can reduce operating costs and is relatively inexpensive and simple to commission.

- Dimming—altering light output with smooth transitions between levels of output—is primarily used for reasons of flexibility in space usage and mood-setting but can also be effective for occupant control of light levels and energy management applications such as daylight harvesting control, scheduled dimming for load shedding and peak shaving, lumen maintenance dimming and night adaptation compensation dimming. Dimming ranges for architectural dimming applications typically range from 100-1% or 100-5%, while dimming ranges for energy management applications typically range from 100-20% (with no significant savings being realized between 20% and OFF due to the fact that efficacy gradually decreases as the lamps are dimmed across the dimming range). It should also be noted that a 10% light level is not perceived as 10% brightness due to the human visual system's ability to adapt. For energy management applications such as daylight harvesting, automatic dimming is more likely to be accepted by users in occupied spaces with stationary tasks. However, it typically poses a higher installed cost and more sophisticated commissioning than switching.

√ What degree of control precision is required in defining the zones? The degree of precision required depends on the number of zones. Each zone constitutes a luminaire or group of luminaires simultaneously controlled by a single device.

- Zones are typically established based on types of tasks to be lighted, use schedules, types of lighting systems, architectural finishes and furnishings, and daylight availability. Zones should be established, if possible, not only on the immediate use of the space but also anticipated future uses. Zones are often limited to the fixtures on one circuit or sub-circuit or switch-leg, but similar fixtures on multiple circuits up to large loads can be zoned for switching by a single controller as well.
- If the goal is to simply shut off an entire building's lighting at a certain time of night, little precision is required, as there is only one zone. If each floor must be switched independently, then a greater degree of precision is required. An example of an application requiring a high degree of precision would be a hotel ballroom. In a ballroom, the lighting is layered as a series of control zones in the same space, resulting in a high degree of flexibility depending on the how the space is partitioned and used.

- The advent of digital lighting systems enables zoning with dimming and switching control at the individual ballast level. A single set of control leads is connected directly from the controls to all of the dimming ballasts, simplifying initial wiring. Additionally, dimming modules and control panels are not needed for digital ballasts to control light output. Ballasts can be controlled individually or grouped into larger zones. As space needs change, the control zones can be reconfigured without changing the wiring.

√ Local or central control?

- In a localized control system (such as wall switches or a task light), each zone is operated by its own point of control independently of other zones, presenting a lower cost and less sophisticated commissioning. In a centralized control system, all zones are operated by a single point of control, providing greater capabilities, flexibility and potential energy savings.
- The control system can be designed with local systems and a centralized system working together as layers. Both local and centralized systems can be integrated into building automation systems for control of lighting and HVAC.

√ What degree of automation is required? The choice depends whether the application calls for the control action to be caused manually by human initiative or automatically according to preset conditions.

- Lighting for visual needs or flexibility of use generally involves the use of manual controls.
- Lighting for energy management needs generally uses automatic controls. For example, occupancy sensors, which activate and/or shut off lights based on detected presence or absence of people in the space, can save significant amounts of energy for a generally moderate cost.[44] Automatic energy management controls are generally more sophisticated but can be more economical because they typically result in higher energy cost savings. However, research indicates that simple manual switching strategies such as bi-level and multi-level switching can also produce cost savings.[45]

√ Controls should be user-friendly, easily accessible (if manual controls) and create lighting conditions that are acceptable to occupants. Programming of the controls should be intuitively obvious, or instructions for their use should be posted inconspicuously on the equipment, so that users do not lose their patience trying to operate them. Maximize user acceptance by educating them about the purpose and use of lighting controls.

Figure 10-4. *This wall has too many switches, which can be confusing to users (and unsightly). (photographer: Naomi Miller)*

√ Avoid excessive switches on the wall; consider consolidating them into a programmable control station.

√ Be cautious about using HID lamps with switches and dimmers. If accidentally switched off, the lamps may take several minutes to re-strike. Dimming is also not instantaneous. The National Electrical Manufacturers Association (NEMA) advises that the lowest recommended dimming level is 50% of rated lamp wattage for both metal halide and high pressure sodium lamps[46]. High-pressure sodium lamps may exhibit flicker during dimming. Metal halide lamps may exhibit color shift during dimming, although this is noticeably reduced with electronic ballasts and ceramic metal halide lamps. In addition, self-extinguishing HID lamps are not recommended for use with dimming systems.

√ Controls should be properly calibrated and control systems should be commissioned in the field to ensure performance that meets design intent. Commissioning involves systematically testing all controls in the building to ensure they provide specified performance, interact properly as a system, and fulfill both the design intent and owner's needs related to user satisfaction and, if applicable, energy savings. Calibration of occupancy sensors and photosensors is an essential step in commissioning systems featuring such controls; designers can consider self-calibrating controls such as self-calibrating occupancy sensors. Even when self-calibrating controls are used, however, commissioning should be planned, budgeted and executed as part of the design and construction process.

Figure 10-5. *User-friendly, clearly organized lighting control. (Photo courtesy of Lutron Electronics.)*

LIGHT + NIGHT

When it is darkest, men see the stars. — Ralph Waldo Emerson

Emerging Outdoor Lighting Issues

At night, people use outdoor lighting to travel, work and play. Nighttime lighting can enhance safety and security, beautify urban landscapes and enable a wide range of visual tasks, from driving a car to watching a live baseball game.

Despite the benefits of outdoor lighting, much of the light that is produced by these lighting systems ends up where it's not intended, from neighboring properties to the sky, which is transformed from a pleasing view of millions of stars to a milky haze over most cities.

This chapter describes how to minimize light pollution (skyglow) and light trespass, while addressing additional nighttime lighting issues including glare and peripheral detection.

Light Pollution/Light Trespass

Light pollution, or *sky glow*, results from light that is emitted upward or reflected from pavement or groundcover upward into the sky from outdoor luminaires.

Light pollution interferes with astronomical observations and obscures a view of the stars for nine out of 10 people in the United States. Migratory birds, sea turtles and other forms of wildlife are negatively affected by nighttime lighting.[47] Light traveling up into the atmosphere or onto neighboring properties is also a waste of money for the owner.

There are many contributing factors to light pollution, including both natural and electric light sources. Examples of electric sources of light pollution include street lighting, wallpacks and porch lights with poor optics that send some light upward rather than directing it to the road or wall surfaces. In addition, some percentage of useful light falling on light-colored parking surfaces, roads and landscape features — particularly when there is snow cover — is reflected to the sky and contributes to skyglow. Even skylights that reduce daytime building energy use can inadvertently transmit electric light into the sky if the building is lighted at night.

Light trespass is light that travels into adjacent properties, causing a nuisance to neighbors and often leading to new local laws after complaints and litigation.

Figure 11-1. (left)
Figure 11-2. (right)
During the 2003 Blackout, people saw millions of stars over big urban centers for the first time in decades. During the blackout, the image on the left was taken (28mm, f2.8, Fuji 800, approximately 90 seconds, driven). After the lights returned, the image on the right was taken (28mm, f2.8, Fuji 800, 30 seconds, tripod-mounted). The contrast illustrates one of the causes of light pollution. (Photos courtesy of © 2004 Todd Carlson.)

Examples include light spilling from a gas station canopy next door into a bedroom window, light from a car dealership at the end of the block, or a next-door neighbor who has mounted a glaring wallpack to light his or her backyard all night. Often, the nuisance is not the amount of light falling on a window, but the offending brightness of the luminaires used, even when viewed from half a kilometer (one-third mile) away. In some cases, the light is distracting not because of its brightness, but because it is switched on and off frequently, causing abrupt and noticeable changes in lighting conditions from an external source.

A growing number of states and local governments have adopted light pollution/light trespass legislation. Light pollution/light trespass prevention is also encouraged in the U.S. Green Building Council's LEED rating system.

Glare and Outdoor Lighting

Even when light is distributed where it is intended in an outdoor lighting application, the installation may be glaring, which can affect visibility and safety. Glare is direct discomfort or disability glare from the luminaire (see Chapter 3 for more information about the causes and effects of glare).

When designing an outdoor lighting system, consider glare as a design criterion that is as critical as illuminances. It is important to remember that in outdoor lighting applications, despite what laypeople may believe to be true, glaring lights do not improve visibility. In fact, glare usually does the opposite. Visibility will improve when glare is reduced.

How to Minimize Light Pollution/Light Trespass and Glare

For more information about exterior and parking lot lighting, light pollution and light trespass, see IESNA RP-8, RP-33, RP-20, TM-10 and TM-11.[48, 49, 50, 51, 52]

- √ Recommend outdoor lighting only where it is really needed. First ask: "Is light really needed here?" The project should be broken into areas where nighttime lighting is appropriate or not appropriate; only light areas where it is appropriate.

- √ Luminaires with excellent optical control and shielded luminaires are usually a better choice.

- √ For both light pollution and glare control reasons, select outdoor luminaires with optics that emit no light (luminous intensity, measured in candelas) at or above 90º above nadir whenever possible, and which limit light emitted at high angles (at or above 80º). (Nadir is 0º, a line pointing straight down from the luminaire to the ground; 90º represents horizontal.) There are some luminaires that deliver almost all of their light downward (98% or more) with excellent glare control. Check photometric information carefully especially if a decorative luminaire is used, because decorative luminaires often have poor optical design that directs too much light at high angles (80° and above).

- √ Lay out the lighting carefully to achieve the recommended light levels and uniformities with a minimum of total lamp lumens. There are many computer programs that can help calculate light levels and this in-

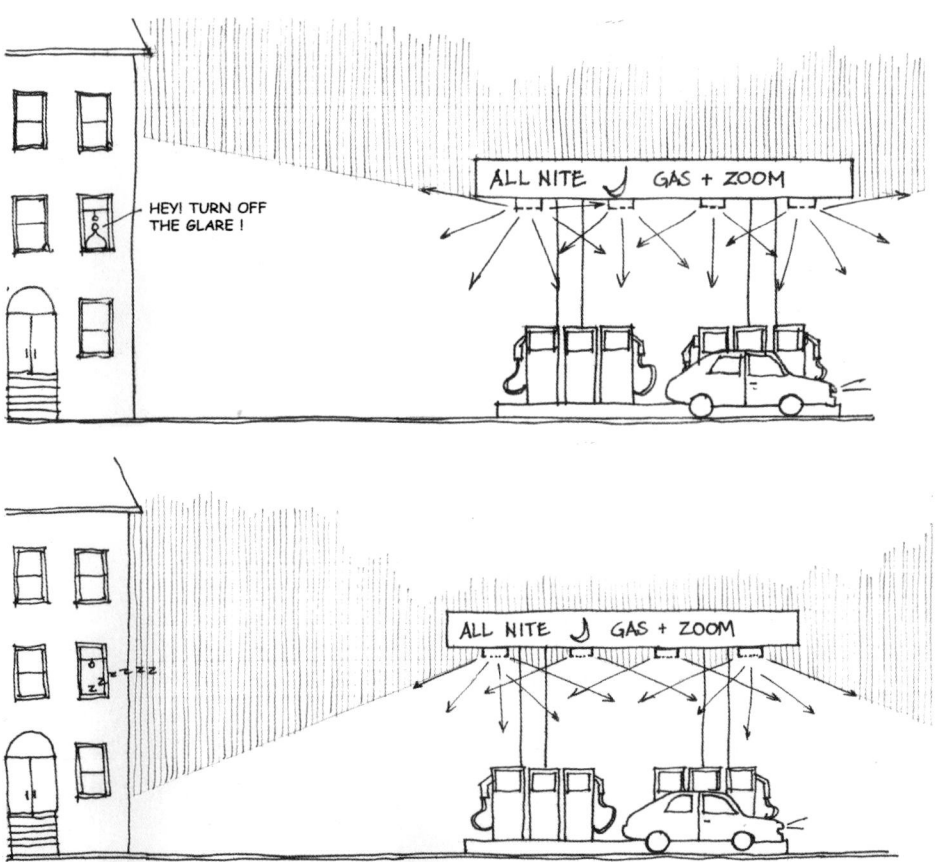

Figure 11-3. *(top)* **Figure 11-4.** *(bottom)*
Gas station canopy luminaires often emit light at high angles, causing glare and light trespass for neighbors. Canopy luminaires with better optics limit luminous intensity distributed above 80°. (Drawings courtesy of Naomi Miller.)

Figure 11-5. *Decorative walkway lighting with optics that direct all light downward. Lincoln Financial Field. (Photo courtesy of Architectural Area Lighting.)*

formation can help the designer to evaluate whether there is a glare condition for drivers or a light trespass issue.

- Be responsible about target illuminances for commercial properties. Some gas stations, for example, have been using excessive illuminances to attract customers—in some cases delivering more than 1,000 lx (100 fc) at the pumps in an effort to make the station appear open, clean and safe. Excessively high light levels, however, can make it harder for the driver's eyes to re-adapt to the much darker street, which can present a significant safety hazard. An impression of openness, cleanliness and safety can be achieved with much lower, uniform and low-glare lighting.

√ Choose lamp wattages and luminaire optics that distribute the right amount of light where it is needed. Use house-side shields, louvers and baffles to reduce spill light from luminaires onto adjacent properties.

√ Use lighting controls to automatically shut off lighting when it is not needed.

- In outdoor areas with limited hours of nighttime use, such as a community sports field with a fixed closing time, shut off unnecessary lighting at a curfew hour (approximately 10 PM) using automatic time-clock controls.
- Similarly, turn off decorative lighting, building floodlighting, landscape lighting and any unnecessary outdoor lighting after 10 PM (or 11 PM or midnight, depending on application and location). This provides astronomers a few hours of observation time every night. Do not rely on staff or homeowners to shut off the lights: install time clocks to do the work automatically.
- In 24-hour retail parking areas, consider shutting off lighting at the perimeter of the parking lot after 9 or 10 PM, when the number of customers drops off and the customers tend to park near the entry doors.

CHAPTER 11: LIGHT + NIGHT

Figure 11-6. *Light trespass into bedroom windows from gas station canopy luminaires.*

(Peripheral lighting can contribute to a perception of safety for employees and customers, however, so evaluate whether shutting off the far perimeter lighting will affect that perception of safety.)

- Alternately, lights can be shut down to half level after the curfew hour if crime is a concern, as long as the resulting light patterns provide needed levels of visibility and still meet uniformity recommendations. Uniformity is an important component of security lighting.

- Motion sensors, used in conjunction with an instant-on light source, can reduce nighttime lighting use when it is not needed. Motion sensors switch off outdoor luminaires when a space is unoccupied and activate the luminaires when the presence of people is detected. Alarm systems and cameras can be connected to sensors to summon security or police and take images of would-be trespassers.

√ Specify porch lights, post-top luminaires, bollards and wallpacks that throw a minimum of light upward.

√ Consider higher mounting heights for sports lighting luminaires. This allows the designer to use narrower beam distributions and keep the aiming angle low (more downward), resulting in less glare and trespass for surrounding properties.

Figure 11-7. *(left)* **Figure 11-8.** *(right)*
Parking lot luminaires with house-side shields. (Photo 11-7 courtesy of SELENE; Photo 11-8 courtesy of Carol Jones.)

√ Select luminaires for billboards and other signage that light the sign downward rather than upward.

√ Use low-pressure sodium lamps judiciously. While low-pressure sodium lamps are one solution for purposes of enhancing astronomical observation of the night sky, their monochromatic color properties and poor optical properties make them inappropriate for many outdoor applications.

√ Some recommend light that is richer in short wavelengths (e.g., blue) for improved visibility for drivers and pedestrians while also allowing a reduction in illuminances compared to light sources that are weak in short wavelengths. Others advise against using sources with significant short-wavelength content because short wavelengths contribute more to light scattering in the atmosphere. As of the time of writing, this debate is still ongoing.

Peripheral Detection and Outdoor Lighting

This design issue is related to light distribution and also color appearance (see Chapter 5) and describes the ability to detect people and objects in the periphery of the field of view. For example, it is critical for drivers to be able to see pedestrians, animals and cross-traffic when they are not immediately in the driver's direction of view. Ensuring adequate light for peripheral detection is also important to create a perception of personal security in outdoor and some interior applications.

SIDEBAR 11-1. Detection and the Driving Task

If another car is heading into the oncoming path of a vehicle, ideally, the driver will be able to see it well in advance so as to avoid a collision. This detection is accomplished via peripheral vision—"seeing out of the corner of your eye."

Under low light levels (below 0.5 fc, or 5 lx), the rods in the eye, light-sensitive cells which are plentiful around the outer edges of retina, play a significant role in peripheral vision, being sensitive to differences in brightness and to changes within the field of view. They enable the eye to detect a moving object more easily than a stationary one. When an object moves against the background, there is a change in the light-dark patterns on the retina; the rods transmit this information to the brain, and movement is perceived.

The object, however, must stand out from the background, so there needs to be light on objects detected at the edges of the visual field in order for them to be seen. This implies that drivers need light not just on the road surface, but on the shoulder and near-shoulder areas of roads as well.

Further, since rods are more sensitive to short-wavelength light (blues) than longer wavelengths (such as yellows), using light sources such as metal halide lamps instead of high-pressure sodium or low-pressure sodium may enhance the ability to detect objects using peripheral vision under low light levels (mesopic levels).

Research suggests that this is the case: At the same pavement illuminance, subjects reacted faster to objects at the edges of the visual field when the objects were lighted with metal halide light than with high-pressure sodium light. The researchers also found that the lower the illuminance level, the bigger the difference in reaction times. The darker the area is, the more important is may be to have some blue in the spectrum of the light source.

This issue, however, is not entirely conclusive and continues to be the subject of research. There is more to be explored in how the aging eye responds to blue light, the perception of glare from short-wavelength light, and other factors.[53]

CHAPTER 11: LIGHT + NIGHT

Figure 11-9. *A shopping mall parking lot with uneven light distribution, leading to a perception of personal insecurity. (photographer: Naomi Miller)*

Figure 11-10. *A shopping mall parking lot with a relatively even light distribution, leading to a greater perception of personal security. (photographer: Naomi Miller)*

Imagine a shopping center parking lot at night where the lighting around the perimeter is uneven and there are deep, shadowy areas. (See Figure 11-9.)

It would be difficult to detect a potential mugger moving in these shadows, and there would not be a clear place of refuge, leading to perception of insecurity.

Most people would feel more secure in the parking lot in Figure 11-10 because the lighting around the perimeter is more even, and peripheral vision can more easily detect movement.

How to Enhance Peripheral Detection

√ If security is an issue, light the perimeter of outdoor spaces as uniformly as possible, particularly vertical surfaces such as building walls, trees, fences and similar surfaces. If necessary, use lighting to mark a clear path of refuge.

√ The most appropriate light source for enhancement of peripheral detection is not always a clear-cut

Figure 11-11. *Because of its blue-rich spectrum, metal halide (Figure 11-11a [left]) may be a better choice than high-pressure sodium lighting (Figure 11-11b [right]) for a high-security parking lot. (Photo a courtesy of Naomi Miller; Photo b courtesy of Nancy Clanton.)*

choice. On-axis (foveal) vision is supported by light sources that emit most of their energy in the yellow-green portion of the visible light spectrum. Off-axis vision is best supported by light sources rich in shorter wavelengths (blue) of the spectrum. If off-axis vision is important at low light levels—i.e., less than 5 lx (0.5 fc)*, consider light sources such as metal halide or fluorescent that provide both yellow-green and shorter wavelengths. Note, however, that lamps with more short-wavelength content may also appear brighter and therefore irritating to nearby residents or even pose negative health consequences.[54,55] In general, use optical systems that keep stray light from entering bedroom windows so that the spectral content of the light source is less of a glare or health issue for local residents. Pay attention to the spectral distribution of the light source for nighttime driving or where pedestrians need to see at a distance.[56]

---

* For luminance values, this is equivalent to 0.3 cd/m², assuming 5 lx (0.5 fc) on an 18% reflectance surface; 18% reflectance is an average reflectance for trees, grass and pavement, or typical cityscape.

PART THREE: LIGHT + ARCHITECTURE

The sun never knew how great it was until it struck the side of a building.

– Louis Kahn

Figure 12-1. *In this public space, lighting was conceptualized as a graphic element that would reinforce the architecture; the result is striking, minimal and architecturally integrated. (Photo courtesy of Elliott Kaufman Photography.)*

LIGHT + ARCHITECTURE

Architecture is the learned game, correct and magnificent, of forms assembled in the light. — Le Corbusier

Lighting for Architecture

The integration of lighting and architecture presents a number of opportunities and challenges related to function and beauty.

Luminaire selection is influenced by its aesthetics and possible psychological meanings; whether the luminaire should stand out, blend in or disappear into the architecture; how well the luminaire is scaled to the space in terms of size, finish and mounting height; and where the luminaires are to be placed. Other key considerations include furniture finish, size and mounting height.

This chapter describes these factors and provides recommendations for improving space and luminaire appearance.

Daylight is another important point of intersection between light and architecture, covered in Chapter 14.

Appearance of Space and Luminaires

Luminaires and the light patterns they produce are an integral part of any design composition and its architectural style. Luminaires, however, carry "baggage." They are architectural elements that convey meaning and evoke period and style, which can be used to the designer's advantage.

For example, fluorescent striplights connote "discount" in big-box retail stores. A lava lamp in a waiting room evokes the spirit of the 1960s. An open-wire low-voltage lighting system raises expectations of a whimsical, contemporary style in a boutique. Galvanized-metal industrial luminaires can convey a rough-edged casual appearance in a teenage clothing store.

Figure 12-2. *Fluorescent striplights lend a budget look to this grocery store.*

Figure 12-3. *Decorative luminaires add a sense of whimsy to this children's room. Chicago, Illinois. (designer: Aaron Mobarak; photographer: Les Borschke; decorative painting: Celeste Coughlin, Asterisk)*

Architectural Integration

Luminaire selection is heavily influenced by the level of architectural integration desired. In some projects, as discussed above, the luminaire is visible and intended to contribute meaning to the viewer's interpretation of the space. In some cases, the sole function of a lighting product is to entertain and distract. In other cases, the luminaire must blend into the architecture with a neutral style, or disappear into the architecture, illuminating a space with no visible luminaire. In restoration projects, the visible luminaires often need to reflect a specific historical period.

Figure 12-4. *Industrial luminaires contribute to the funky, rough-edged aesthetic in this teenage clothing store. (lighting designer and photographer: James R. Benya)*

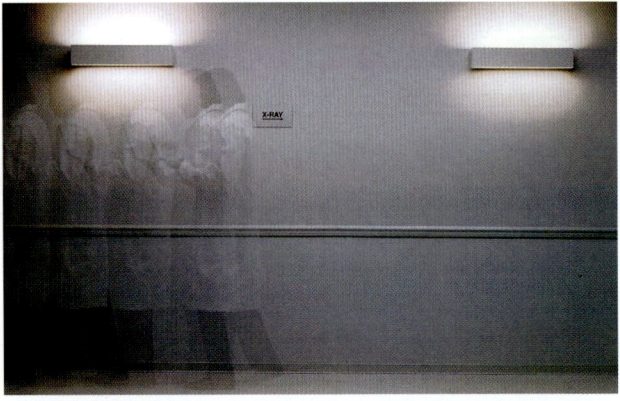

Figure 12-5. *Sometimes, it's appropriate to use "plain vanilla"-styled luminaires that blend with the architecture. (Photo courtesy of Litecontrol.)*

Size, Finish, Mounting Height

Even if a suitable style and level of integration are achieved, the luminaire should be properly scaled and matched to the space. The size, finish and mounting height of luminaires must be carefully considered.

For example, consider a grand two-story apartment building lobby with a small five-armed brass chandelier mounted from the ceiling. In this case, the chandelier, likely selected from a catalog of residential dining table pendants, appears out of scale. Now consider this same space where every metal finish is chrome or polished stainless steel. The brass chandelier appears even more dissonant.

Location of Luminaires

The physical location of luminaires, including where they are located and their locations in relation to each other, impacts the space.

For example, regular grids of luminaires can appear regimented and business-like. A random pattern, as long as it does not result in visual confusion, can convey a more casual atmosphere. Long rows of linear luminaires can visually lengthen a hallway, while recessed wall accent lighting on alternating sides can shorten it.

Locating luminaires requires careful coordination with beams and columns and architectural features.

While a lighting layout may look suitable in the reflected ceiling plan, it may not be the most appropriate solution for the space. Lighting is not experienced "in plan," but viewed in perspective, often with the viewer in motion.

The location of luminaires in relation to each other is also important. For example, when outdoor pole-mounted luminaires are located close to each other, they can create visual noise and affect space appearance. When only light level and uniformity on the pavement is considered, there is a risk of creating a distracting forest of poles and fixtures. Sometimes short poles would work aesthetically better than tall poles, but would necessitate spacing them closer together if the uniformity targets remain the same. Uniformity, light levels and visual clutter need to be examined together to determine the balance point that solves all of the issues well.

Figure 12-6. *Uplights integrated into the cornices of this cathedral highlight the sculptural ceiling with no visible lighting equipment. (Photo courtesy of Rambusch Lighting.)*

Figure 12-7. *A randomized pattern of downlights in the ceiling gives this office cafeteria a more casual look than would be achieved with a rigid grid of downlights. Lighting Research Center DELTA Program. (lighting designer: Gary Steffy; photographer: Robert J. Eovaldi)*

Furniture Finishes and Sizes

Furniture sizes and surface finishes can affect light levels and the distribution of light in a space. If a space is being reconfigured, the lighting should also be reconsidered. For example, a tall workstation partition absorbs much more light than a shorter one. Taller partitions may also shadow more work surface area, so it may be necessary to know where the partition is located relative to the luminaire, and what the spread of light from the luminaire is, in order to ensure that some employees won't be left in the dark.[57]

Tradeoffs

There are always advantages and disadvantages, interactions and tradeoffs when considering options among building systems.

For example, lower furniture panels improve daylight penetration from office windows, uniformity of illumination and access to a view, but can negatively impact acoustic and visual privacy. If privacy is paramount to a client, therefore, the designer may be doing the client a disservice by recommending lower panels even though these panels may improve lighting quality and energy efficiency.

How to Improve the Appearance of Space and Luminaires

√ A successful lighting design supports the architect's artistic intent. Lighting can be used to "paint" a space with light to meet the architect's goals of inspiration, spaciousness, aesthetics, interest, etc.

√ Choose luminaires with a stylistic appearance that supports the space's intended appearance, atmosphere and meaning.

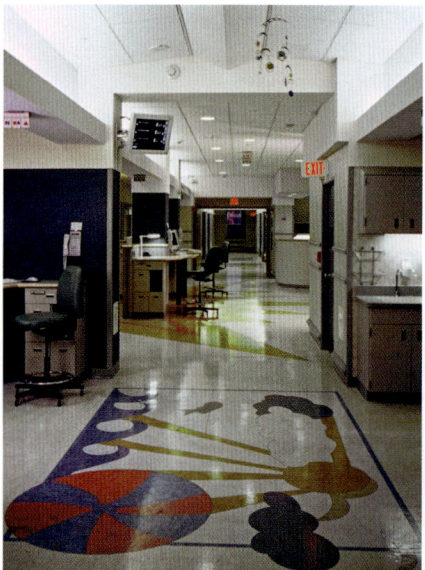

Figure 12-8. *Cove lighting offers one way to integrate luminaires into the architecture. (lighting designer: Naomi Miller; photographer: Mick Hales)*

Figure 12-9. *Ceiling and wall slots can deliver light to the space without visually adding luminaires. (Photo courtesy of Litecontrol.)*

CHAPTER 12: LIGHT + ARCHITECTURE

Figure 12-10. *In this high-end restaurant, some luminaires stand out as distracting and entertaining works of art, while others are integrated with the architecture—the result is the eye is drawn to the illuminated architectural features and the decorative luminaires themselves. Candles add a layer of playful illumination and contribute pools of intimacy. Koi Restaurant, New York. (lighting design: Focus Lighting; architectural design: iCrave; photographer: © 2008 Frank Oudeman)*

√ Carefully choose the luminaires' size, finishes and mounting height to support the design intent.

√ In spaces where the style of the lighting needs to be subordinate to the architectural design, integrate the luminaires with the architecture. Coves, wall slots, ceiling slots, soffits and valances provide ways to conceal luminaires from direct view. If recessed downlights are required, consider small-aperture units.

√ In spaces where luminaires must be exposed to view, but should "blend into the background," choose luminaires that have a neutral style.

√ Locate luminaires so that they respect the rhythm, scale and location of architectural features, but don't sacrifice the resulting lighting effect in the sole interest of conforming to a pattern.

Figure 12-11. *The uplighting luminaires concealed at the edge of this ceiling streak toward the oculus, bringing out the beauty of the patterned wood ceiling. The number and location of uplights reinforces the rhythm of this 16-sided worship space. (lighting design: Naomi Miller; photographer: © Randall Perry Photography, LLC)*

Figure 13-1. *Light patterns can stimulate visual interest. Social House Restaurant. (lighting design: Focus Lighting; architectural design: Avroko; photography: © 2008 Frank Oudeman)*

LIGHT + DISTRIBUTION

I knew, of course, that trees and plants had roots, stems, bark, branches and foliage that reached up toward the light. But I was coming to realize that the real magician was light itself. — Edward Steichen

Patterns of Light

Patterns of light on walls, ceilings and other surfaces can affect how bright and large a space appears, and can dramatically influence user judgment of the space. Even the softness of the edge of a light pattern (the gradient) and where the pattern occurs in the field of view, can affect psychological response to a space.

Light patterns can also stimulate visual interest. For example, light patterns can attract or guide viewer attention by creating points of interest which can be presented individually or within a visual hierarchy. As another example, light patterns can create points of high brightness, or sparkle.

In many workspaces, the pattern of light at the workplane should be relatively uniform so that shadows do not obscure important details. Additionally, the lighting design community increasingly recommends that room surfaces within the occupants' field of view—such as walls and ceiling—be illuminated, which can positively affect the occupants' satisfaction with the space.

The distribution of illumination in a space involves critical design decisions based on how the space will be used. This chapter describes light patterns and how they can be used to attract and guide attention, including how to achieve the benefits of sparkle in spaces that warrant it; uniformity and the steps to ensuring it; and the importance of room surface brightness and how it is achieved.

Light Distribution on Surfaces (Patterns)/ Points of Interest

Light patterns can affect perception of atmosphere and image of a space.

Figure 13-2 presents a romantic restaurant, in which intimacy is more important than task visibility. Average illuminance is low and the lighting distribution is non-uniform. Ambient illuminance is sufficient to find the table, task illuminance is sufficient to read the menu, and accent illumination

Figure 13-2. *Cozy, dark restaurant. (lighting designer: Luminae Souter Associates, LLC)*

Figure 13-3. *Brighter, less intimate restaurant. (lighting designer: Naomi Miller; interior designer: Laura Seccombe; photographer: © Chas McGrath)*

highlights the artwork and interior design. The formless pools of light and the dark walls and ceiling make a dining couple feel less visible, as though they are in a private space.

In Figure 13-3, an equally attractive restaurant is presented, but with different lighting goals. Illumination is more uniform, and the walls are lighter in color and have a higher reflectance, resulting in a more public atmosphere. In this space, task visibility is increased, but the atmosphere is less intimate.

Light patterns can affect the perceived size of a space.

Figure 13-4 shows a corridor with dark walls, which makes the space appear somewhat closed-in. Figure 13-5 shows a corridor with illuminated walls, which make it appear wider. Dark walls make spaces feel smaller, while light walls and ceilings appear to expand space.

Where a light pattern occurs in the field of view may affect perception of brightness; the sharpness of the edge of the pattern may also affect brightness perception.

In Figure 13-6, recessed luminaires with diffuse-finish parabolic louvers distribute high, soft-edge scallops of light on the wall. As a result of the pattern occurring high on the wall and featuring a soft edge, the space appears brighter.

In Figure 13-7, small-cell parabolic louver panels replaced flat prismatic lenses to reduce computer screen reflections. While this goal was accomplished, the luminaire's

Figure 13-4. *Dark walls make this corridor feel like a tunnel. Lighting Research Center. (photographer: Russ Leslie)*

Figure 13-5. *Lighted walls make this corridor appear wider than its actual measurements. (Photo courtesy of Litecontrol.)*

CHAPTER 13: LIGHT + DISTRIBUTION

Figure 13-6. *High, soft-edge scallops of light on the wall from recessed luminaires with diffuse-finish parabolic louvers make this space appear bright and cheerful. (Photographer: Cindy Foor, Focus Studios; Photo courtesy of Lighting Research Center.)*

Figure 13-7. *In this installation, small-cell parabolic louver panels replaced flat prismatic lenses to reduce computer screen glare. This strategy accomplished the goal, but the luminaire's resulting narrow pattern of light creates a gloomy, cave-like atmosphere because the top part of the wall is dark and the edge of the light scallop is sharp. (Photo courtesy of James Benya, Benya Lighting Design.)*

resulting narrow pattern—lower on the wall, and with the edge of the light scallop being sharp instead of soft—creates a space that appears cave-like in comparison.

Light patterns can help draw or focus attention on the most important elements in a space, called *points of interest*.

Lighting can be used to highlight key space features using accent lighting or wallwashing. Figure 13-8 shows the lighting of an important focal point in a house of worship. Alternatively, an energetic luminaire itself, such as a decorative luminaire or light sculpture, can serve as a central point of interest, as shown in Figure 13-9.

Light patterns can support the architect or interior designer's vision, reinforcing regular patterns of wall panels or bookshelves. Patterns that differ in intensity can also help create a visual hierarchy in the environment.

Figure 13-8. *The lighted sculpture of the Holy Family provides an important focal point in this Reservation Chapel. (lighting designer: Naomi Miller; photography: © Randall Perry Photography; James Hundt, Architect)*

Figure 13-9. *Chicago O'Hare International Airport (Photo courtesy of Sergio Mazon.)*

Figure 13-10.

Figure 13-11. *Examples of how light patterns can support a visual hierarchy. (13.11 lighting design: Naomi Miller and James Benya; photograper: Ross de Alessi.)*

Figure 13-12 and 13-13. *The spots of light on the sign and jewelry case don't correspond with the information on the sign or in the case. (Photos courtesy of Naomi Miller.)*

CHAPTER 13: LIGHT + DISTRIBUTION

Figure 13-14. *Hotel entry canopy with multiple light patterns that can distract attention from the curb. Seattle World's Fair; John Flynn, photographer.*

Figure 13-15. *The light pattern on this train station platform, in combination with the paint color, clearly marks the platform's dangerous edge. (Photo courtesy of the Chicago Department of Transportation; lighting designer: Schuler Shook.)*

In the store window in Figure 13-10, the viewer's eye is drawn to the center grouping of china, where there are three lights, then to the displays on either side, where there is only one light. In Figure 13-11, which presents St. Mary's Cathedral, the eye is first drawn to the worship leaders and the altar, then to the shimmering Baldacchino above, then to the cherry wood walls behind.

Light patterns must make sense or else be potentially distracting.

In Figure 13-12, the spot of light on the sign does not correspond with the information on the sign. In Figure 13-13, the spots of light in the jewelry case do not correspond with the presentation of the jewelry, actually visually reducing the importance of the merchandise. This can lead to confusion, or "cognitive dissonance," for the viewer.

Light patterns can distract from important information in a scene, particularly concerning safety.

In Figure 13-14, the random patterns on the ground beneath this hotel entry canopy mask the curb; the distraction may cause guests to trip over it. In Figure 13-15, the lighting on this London train station platform, coupled with a different paint color, clearly marks the platform's edge.

How to Achieve Appropriate Light Patterns on Surfaces and Objects

√ The designer should visualize the three-dimensional space under design, and identify important features. These features can be prioritized in visual importance, based on what should be noticed most, and which features, if any, should be concealed. This will determine the hierarchy of what should be lighted.

- √ Identify lighting needs based on the intended use of the space, if it should appear public or private, larger or smaller, and whether the lighting should be uniform or varied.
- √ Light patterns should make sense, corresponding to artwork locations, rhythm of wall panels, signage locations, important architectural elements and similar features.
- √ Increase illuminances on work surfaces and artwork if the viewer's eye should be drawn to them, and place emphasis on those areas. This may increase user acceptance.[58]
- √ Avoid harsh or striated patterns, especially when the pattern falls low on workplace walls.
- √ It is often appropriate to light workspaces with a more uniform distribution on walls and ceilings compared to other space types. Luminaires and light patterns that are organized and harmonious can project an atmosphere of calm without being dull.

Sparkle (or Desirable Reflected Highlights)

High levels of "brightness," the perceived response to luminances in a space, can be irritating, even intolerable (see Chapter 3). Yet if exploited correctly, it can be a pleasing and desirable highlight. This is when brightness is transformed into sparkle.

Sparkle typically presents itself as a tiny and bright point of light shining alone or in clusters. Sparkle and glare are easily distinguished; if a point of brightness becomes too large in area or too bright such that it causes discomfort or impairs vision, it is glare and should be avoided or removed. Conversely, if the tiny dot of light loses its brilliance, the sparkle loses its scintillation.[59]

Research indicates that sparkle is achieved when luminance is above 10,000 cd/m^2 and the size of the source of luminance is smaller than 10^{-6} steradian—smaller than the size of Lincoln's head on a penny viewed at 5 meters, or about 16 feet, away.

Sparkle has been used to create highlights and visual interest in a space when warranted by the application.[60,61,62,63] Popular examples include chandeliers, candles or light playing on highly specular surfaces such as silverware. When properly integrated in a lighting design in which highlight and visual interest is required, sparkle can ignite the spirit of the space and express excitement, high quality or elegance. For example, sparkle can play a significant role in some retail displays such as glassware or china, and can also add a touch of elegance to a hotel lobby, restaurant or living room. If designed properly, sparkle can add an uplifting element even in more serious environments such as offices.

In some cases, sparkle can play an important functional role; in an industrial application, for example, these reflected highlights may be critical for checking the quality of the paint finish in an automobile paint line or the quality of finished metalwork in a chandelier factory.

Figure 13-16. *If designed properly, sparkle on luminaires can uplift serious environments such as offices. (Photo courtesy of Peerless Lighting; photographer: © 2004 John Sutton Photography.)*

The sensation of sparkle is affected by three factors—the intensity of the luminance, the visual size of the luminance, and the light level in the space.

How to Create Sparkle

- √ Use point sources such as clear incandescent or metal halide lamps or LEDs shining through punched holes in a decorative wall sconce or pendant.

- √ In chandeliers, expose the lamps and use low-wattage clear lamps so that the filament becomes the source of sparkle without becoming glaring.

- √ Highlight glass objects or highly reflective surfaces with numerous small point sources.

- √ It is harder to achieve sparkle when using fluorescent or coated metal halide lamps because the size of the source is too large and uniform in brightness. However, sparkle can be created with these lamps by allowing the lamp brightness to pass through small punched slots or holes. These small openings themselves serve as the sources of sparkle.

Figure 13-17. (Left) *This restaurant booth exhibits sparkle using a simple punched metal shroud wrapped around clear incandescent lamps. (lighting design: Luminae Souter, Associates LLC)*

Figure 13-18. (Right) *Low-wattage, clear incandescent lamps, coupled with sparkling crystal glass elements, help this senior residence entry convey elegance. (lighting design: Naomi Miller; interior design: Laura Seccombe; photography: © Chas McGrath)*

Figure 13-19. *Tiny, bright points of light inside the display cases makes the gems sparkle. (Photo courtesy of Visual Lighting Technologies.)*

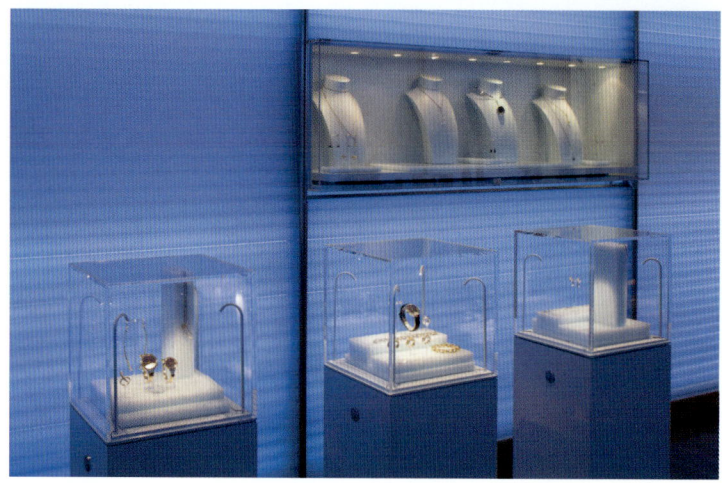

Light Distribution on Task Plane (Uniformity)

Uniformity, the even distribution of light intensity across the task plane, is desirable in many workplace applications. The *task plane* or *workplane* is a surface, real or imaginary, where visual work takes place. This is where the designer is concerned about providing sufficient illuminance. The plane may be horizontal, vertical or slanted; a given space may have several task planes.

In a library stack area, the task plane is the vertical surface where people read and select book titles. In an office, it is usually the horizontal surface at the desktop level, or about 76 centimeters (30 inches) above the floor. In a covered parking garage, it is the concrete floor and the walls. In a museum, it may be a painting hung on the wall. In a classroom, the task planes may be the chalkboard or whiteboard, the face of the teacher, and the student desktops.

When illuminance is not uniform, task visibility may be affected. In a library stack area, for example, it may be easy to find book titles in the brightly lighted areas, but not in the shadows; a library patron may find that she can still locate the book, but that it takes her longer. Experience has shown that keeping illuminances within a limited range helps maintain visibility across the task plane. That is why library stack lighting, desktop lighting, industrial lighting, parking lot lighting, etc. is often recommended with a uniformity ratio. Depending on the application, the uniformity ratio may be recommended to be less than a certain maximum-to-minimum illuminance ratio, or an average-to-minimum illuminance ratio.

Figure 13-20. *Overhead downlights produce poor uniformity on this workstation desktop.*

How to Achieve Good Uniformity on the Task Plane

√ Follow guidance from the IESNA on maximum-to-minimum illuminance ratios to achieve uniformity.

√ Identify the task plane and consult the latest edition of the IESNA *Lighting Handbook* application chapters or the latest IES Recommended Practice publications to determine if there are task plane uniformity guidelines for the application. Depending on the application type, these may be expressed as illuminance maximum-to-minimum uniformity ratios or average-to-minimum ratios. For example, as overhead cabinets in a workstation can cast shadows—making areas at the rear of the desk appear dark—it is recommended to maintain illuminance in the working area of the desktop within a 2:1 maximum-to-minimum ratio[64], which also means ensuring that task lighting not be dramatically higher than the surrounding lighting on the desk. In an outdoor parking lot, dark areas can obscure curbs and obstacles; IES recommends no more than a 20:1 maximum-to-minimum ratio on the parking surface (IESNA *Lighting Handbook*, Ninth Edition, Table 22-21).[65]

√ After determining the basic lighting layout, do a mockup or perform a point-by-point illuminance calculation to identify problem areas. If there are any places where illuminances are higher or lower than the recommended range, consider changing the luminaire and its light distribution, or experiment with a different mounting height or luminaire layout to ensure good uniformity.

√ Non-uniform illuminances matter more when the patches of light have sharp or abrupt edges than when the patches of light have a softer edge (or gradient) on a surface. For example, a 10:1 maximum-to-minimum luminance ratio on an office ceiling may be almost imperceptible if the transition from light to dim on the ceiling is gradual and takes place over a large area.

√ For warehouse lighting, library stack lighting and other tall and narrow areas, light-colored flooring can dramatically help improve lighting uniformity at the bottom of the rack or stack. In fact, light colored room surfaces and indirect lighting helps improve lighting uniformity in many types of spaces.

√ Consider adding task lighting to boost low illuminance values and improve uniformity. Because the light is so close to the task, a low-wattage task light can be an energy-efficient way to add helpful light to a small area.

SIDEBAR 13-1. Importance of Task Plane Uniformity in Industrial Applications

An industrial space typically features machinery, some of it hazardous—cutting, chopping, punching, pounding. The location of machinery, and the layout of tasks and lines, may change frequently. Further, the task plane in an industrial space often is not horizontal, but instead vertical or some angle in between. People may work with objects of many sizes and shapes, and the machinery and materials may cast shadows on an important part of the work area.

In these spaces, because tasks and machinery may change, it is often desirable to deliver uniform illuminance on the task plane over the entire work area. Luminaires that provide a broad distribution of light can help direct the needed lumens into niches, making illuminance on the task plane more uniform, improving task visibility and making the space safer for work.

The owner should be encouraged to paint the overhead area, walls and, in some cases, the floor a lighter color, if possible, to increase the amount of light that arrives at any given point from many angles. This produces diffuse light, reducing shadows while creating a brighter environment. In addition, the luminaires will appear less bright against a light-colored ceiling than against a dark background, thereby improving visual comfort.

To maximize the utility of a light-colored ceiling, luminaires may be selected with up to 40% uplight, depending on overhead reflectances, obstructions and the amount of dirt and dust in the environment. With these luminaires, at any given location in the space, light will strike the task plane directly from the luminaires and indirectly, as reflected light, from other directions. In dirtier environments or those with lower reflectances or a lot of overhead clutter, the uplight component can be limited to 15%. Even that small amount of uplight, however, can reduce shadows and make the space appear brighter.

Room Surface Brightness (and Surface Characteristics)

Consider driving in a tunnel where only the roadway is lighted. Now consider driving in a tunnel where the walls and ceiling are illuminated with a wash of light. The first tunnel appears somewhat claustrophobic, threatening—the second tunnel less so.

The same principle appears to be true in interior building lighting.

Conventional wisdom once emphasized horizontal illuminance as the primary metric of success in office lighting design. Designers, however, have long held the view that room surfaces play an important part in occupant perception

Figure 13-21. *(left)* **Figure 13-22.** *(right)*
A tunnel, before and after lighting. The entrance appears much less forbidding when there is some wall surface brightness. (lighting design: Paul Lutkevich, Parson Brinkerhoff; Photo courtesy of Ed Morel.)

Figure 13-23.

Figure 13-24.

Figure 13-25.

Figure 13-26.

Figure 13-23–Figure 13-26. *Photometrically accurate computer renderings of the same office using different lighting solutions- lensed troffers (top left), downlights (top right), parabolic troffers (bottom left) and linear indirects (bottom right). Each solution provides the same average illuminance at desk height. (Computer renderings courtesy of Leslie North.)*

and mood. This belief has received preliminary confirmation by research.

A recent study in a realistic office laboratory in which participants had dimming control over various luminaires suggests that people prefer to have light on vertical surfaces in the field of view.[66, 67] The results suggest 30 cd/m² as a minimum preferred luminance for cubicle panels.

Other studies have derived similar values for other types of office walls.[68] A similar conclusion was reached in studies of people viewing digital projections of simulated office spaces.[69] As an example, 200 lux (20 fc) falling onto a diffuse light gray surface (50% reflectance) produces a luminance of about 30 cd/m².

Additional reasons why surface brightness within the field of view, sometimes called *volumetric brightness*, is desirable in workspaces:

- Light-colored surfaces reflect diffuse (multidirectional) light back into the room, increasing the overall effectiveness of the lighting system.
- The multiple directions of reflected light help wash out distracting shadows on work surfaces.
- Bright surfaces provide soft, horizontally directed light, which helps faces appear more pleasant and easier to identify and see.
- Bright surfaces mitigate veiling reflections of shiny printed material.

Figure 13-27. *Light reflected from ceilings and walls provides vertical illuminance on faces and objects, making them more visible and pleasant in appearance. (Drawing courtesy of Yukiko Yoshida from original sketch by Naomi Miller)*

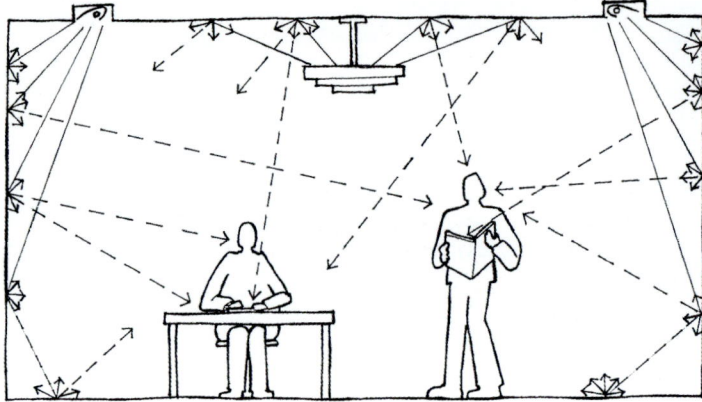

- Bright surfaces increase the adaptation luminance of users in the space, which means the users are likely to see more easily and experience less glare.[70, 71]
- Bright surfaces may even save energy, as volumetric brightness can make the space appear brighter without delivering more illuminance to the task plane.

Room surface brightness, however, is not always desirable, as it may also connote a businesslike, public atmosphere, which would not be appropriate for applications where an atmosphere of intimacy is being achieved. For example, a wine bar is a space where a more private, subdued atmosphere might be appropriate. In this case, it may be advantageous to keep the walls and ceiling surfaces darker, except for bits of accent light to add visual interest.

Figure 13-28. *Recessed troffers with parabolic louvers result in darkness at the top of the walls and scallops between the luminaires. (Photo courtesy of Lithonia Lighting.)*

Figure 13-29. *Recessed troffers with low-brightness diffusing elements provide an indirect component that results in even illumination almost all the way to the top of the walls. (Photo courtesy of Lithonia Lighting.)*

SIDEBAR 13-2. Reflectance/Reflectances

Colors and finishes of room surfaces can have a significant impact on the quality of the luminous environment, as different colors and finishes reflect light in different ways.

Reflectance has two aspects. The first is the *reflectance value*, which describes the percentage of light that is reflected from the surface instead of absorbed. In lighting calculations, this value is represented by the Greek letter *rho* (ρ).

Light-colored materials have a higher reflectance, reflecting more light and absorbing less light (and heat), than dark-colored materials.

The second aspect of reflectance of room surfaces is where they fall in the range from shiny to matte—i.e., how *specular* or *diffuse*. Shiny surfaces do not necessarily reflect more light than matte surfaces; they simply reflect light in a different manner.

Shiny materials can create delightful glints and reflections that are desirable in an entertainment space, but such specular reflections can be distracting, annoying or even dangerous in an office, school or factory space. Shiny surfaces will show a clear lamp (or luminaire or window) image depending on the direction of view, but otherwise appear dark. A matte surface will reflect the light in all directions, creating a softer appearance and contributing to a more uniform illumination; if the matte surface is light in color, it will make the space appear brighter. Glossy, semi-specular surfaces will exhibit both directional and diffuse characteristics, depending how close they are to either extreme. In most cases, it is important to avoid glossy finishes, except as occasional accent surfaces.

High-reflectance room surfaces appear brighter and provide useful inter-reflections in the space that soften shadows and improve comfort with both electric lighting and daylighting. Higher room surface reflectances can also play a role in energy-efficient lighting design because less light is needed for both ambient and task illumination.

High-reflectance surfaces—such as white-painted or very light tints of white, 70-90% reflectance—are necessary in daylighted spaces, especially on ceilings and the window wall, to reduce contrast and window glare and enhance interreflections.

Dark-colored surfaces—such as deep blue paints or dark wood tones, which are typically 5-15% reflectance—can create uncomfortable extremes of brightness for occupants while wasting energy by absorbing light. The use of dark mahogany walls in a conference room, for example, can make it impossible to achieve an acceptable balance of brightness; lighter-stained wood used as wainscoting and trim, coupled with higher-reflectance materials for walls and ceiling, would be preferable for visual comfort. Sometimes, occupants will even complain about not having enough light when their walls are dark in color, even though their workplane illuminance is more than the recommended level.

Even when intentionally creating a darker environment, such as an intimate restaurant, it is preferable to use medium-reflectance finishes—such as medium blues, grays or reds at 20-40% reflectance—and direct the light away from those surfaces, rather than use very dark finishes.

In some projects, it is advisable for the architect or interior designer to include a column of recommended reflectances on the finish schedule so that the entire design team is aware of the selections. These values are available from fabric and paint suppliers.

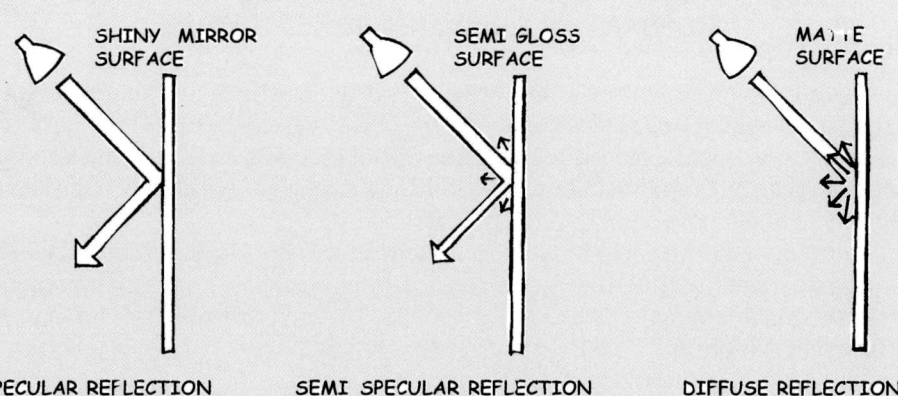

Figure 13-30. *Specular reflection, semi-specular reflection and diffuse reflection. (Drawing courtesy of Yukiko Yoshido from original sketch by Naomi Miller.)*

Figure 13-31. *(left)* **Figure 13-32.** *(right) The first (Figure 13-31) and second (Figure 13-32) floors of this healthcare providers facility have almost identical illuminances on the floor, desktop and ceiling. However, the first floor space appears dimmer because the reflectance and therefore the luminance of the dark green cabinets are lower. (lighting designer: Gary Steffy; photographer: © Robert Eovaldi)*

How to Achieve Room Surface Brightness

√ Surfaces within the field of view should have a high luminance. This can be accomplished with high-reflectance acoustical ceiling tiles, lighter paint colors and washes of light on ceilings and walls.

√ Surfaces do not have to be white; reflectances of 70% and higher are available in many pastel colors. Darker or more deeply saturated colors can be used as limited accent bands, or on floors, chairs, furniture or wainscoting below the height of the work surface.

√ If ceilings are dark, cluttered or low, and light cannot be easily distributed onto the ceiling, increase wall luminance with wallwashing, accent lighting, wall sconces and similar strategies.

√ Besides room surfaces, workstation panels, desktops, shelving and cabinets in the field of view should also have high luminances. One way to achieve this is to increase the reflectance of these surfaces by selecting light-colored fabric, paint and laminates.

SIDEBAR 13-3. Luminance Versus Exitance

When designing for room surface brightness, it is often useful to be able to determine the amount of light reflected by a surface towards the viewer's eye. This takes the specularity of the ceiling or wall surface into account because the luminance of a patch of wall depends on the viewing angle, the wall's shininess and whether the viewer is seeing the mirror-like reflection of the luminaire or window on that patch of wall.

Luminance on a specific area of wall, for example, can be calculated or measured in candelas per square meter (cd/m^2). One may also encounter an obsolete English system unit called "footlamberts" (fL) in luminaire photometric reports. To convert fL to cd/m^2 (on a perfectly diffuse surface only), multiply by 3.14.

This provides information about how "bright" the wall is.

It can also be useful to understand the average amount of light reflected from a wall area irrespective of the specularity of the surface or direction of the reflected light. For this purpose, exitance, in units of lumens per square foot or meter, can be calculated. Exitance indicates the total amount of light reflected from a surface.

The relationship between exitance and illuminance is expressed as:

$$M = \rho * E$$

Where M is luminous exitance in lm/m^2 or lm/ft^2;
ρ is luminous reflectance expressed as a percentage; and
E is illuminance on the surface, expressed in lux or footcandles.

For example, consider a patch of wall finished with flat white paint (0.85 reflectance) and illuminated at 1000 lx (100 fc). The luminous exitance (M) in this case is 0.85 * 1000 lx, or 850 lm/m^2 (0.85 * 100, or 85 lm/ft^2). The "brightness," which is the perception of luminance, has no number but may be judged high or low compared to other surfaces in the space.

The relationship between luminance, exitance and illuminance is expressed, assuming a matte or Lambertian surface, as:

$$L = M / \pi$$

Or:

$$L = \rho E / \pi$$

... where L is luminance in cd/m^2 or cd/ft^2.

Figure 14-1. *Skylights were added to the roof to maximize daylight penetration into this office and casual meeting/conference space. (lighting designer: Naomi Miller; photographer: Gary Hall Photography, Burlington, VT)*

LIGHT + DAYLIGHT

The history of architecture is the history of man's struggle for light — the history of the window. — Mies van der Rohe

Daylight and View

Daylight has served as the primary source of light throughout human history, and has enhanced the most beautiful architectural spaces. In recent years, design for daylight has become a more important feature of mainstream construction in North America due to the sustainable design movement. Daylighting is the use of daylight as a significant source of general illumination in a space.

For the purpose of this publication, "daylighting" refers to the use of diffuse daylight rather than direct sunlight penetration, which can cause extreme luminance contrast and thermal impacts on occupants. Diffuse daylight may come directly from the sky, or from sunlight or skylight that is reflected off of external objects, such as the ground, neighboring buildings and light shelves.

Numerous studies over the last 50 years attest to the importance of daylight in design. Research suggests that daylight can improve user satisfaction and performance, student learning rates and retail sales.[72, 73, 74]

Research also suggests that users are more satisfied when they have a view of the outdoors, and prefer spaces where view windows constitute a minimum of 20-25% of the perimeter walls.[75, 76] These characteristics can make daylighted buildings more valuable and marketable. Daylighting also enables daylight harvesting, an innovative control strategy that can generate 35-60+% energy savings by reducing electric lighting use when ample daylight is present. A daylight harvesting system decreases electric light contribution as the daylight contribution increases.

Advantages of daylighting include:

- High light levels that can make room surfaces brighter and tasks easier to see
- An abundant, natural, renewable resource that can result in significant energy savings
- A view and connection to the outdoors when windows are used
- Excellent color rendering

- Opportunities to relax the eyes at view windows by focusing at a distance
- Interesting dynamic natural variations
- Opportunities to reduce energy costs via daylight harvesting control strategies such as automatic or manual switching and dimming
- Rich in wavelengths that help maintain human circadian rhythms and immune system health. Direct exposure allows the skin to absorb ultraviolet radiation that helps the body synthesize Vitamin D, which is in turn critical for bone health.[77,78] (However, ultraviolet wavelengths are largely absorbed by glass, plastic and architectural surfaces, so occupants normally need to get outside to receive the Vitamin D benefits of daylight).

Figure 14-2. *Daylight provides interesting dynamic natural variations and a sense of connection with the outdoors. "Casa Nueva." (architect: Blackbird Architects; photography: © Bill Dewey)*

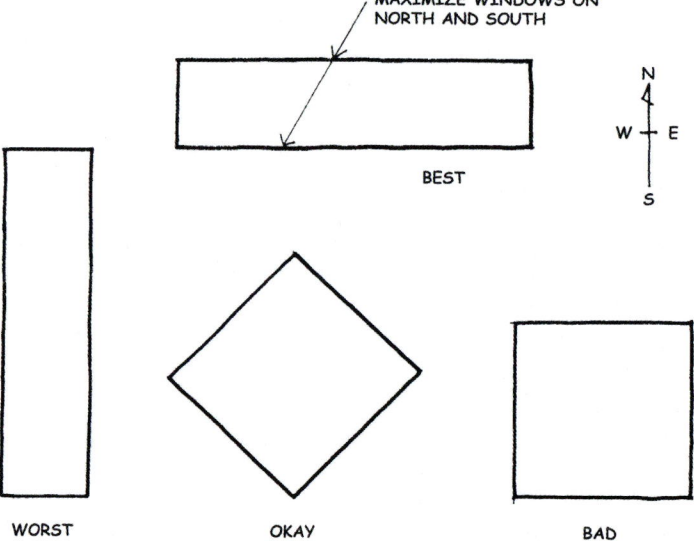

Figure 14-3. *Building footprints with recommended window orientation for northern latitudes. (Drawing courtesy of Yukiko Yoshida from original drawing by Hayden McKay)*

CHAPTER 14: LIGHT + DAYLIGHT

Figure 14-4. *Direct daylight can reduce comfort and impair vision.*

Figure 14-5. *External building feature such as overhangs, fins and awnings can limit direct daylight penetration while increasing indirect or reflected daylight penetration. In this project, an external trellis with Wisteria vines provides additional shading and a pleasing point of integration with ecology. "Casa Nueva." (architect: Blackbird Architects; photography: © Bill Dewey)*

Challenges include:

- High luminances which may cause excessive contrast and glare
- Requirement for solar control and glare control
- Thermal gains and losses that can decrease thermal comfort and increase energy consumption
- Distracting dynamic luminance variations and patterns
- Loss of privacy due to greater window area

Daylighting design requires close coordination between architectural, interior space and lighting design, including design decisions related to building orientation and shape, mechanical system loads, ceiling heights, the design of windows and daylight control, interior configuration and finishes, lighting and control design, and proper calibration and commissioning of controls. As such, it requires diligence, expertise and commitment across the design team.

How to Avoid Glare

Daylight can provide much higher illuminances and luminances than are typically found or needed in interior spaces; higher luminances, and extremes in contrast between luminances in adjacent visual areas, can result in glare.

Uncontrolled direct daylight penetration can create luminance ratios of more than 1000:1, which is unacceptable for most visual tasks. In a workspace, direct daylight can create excessive contrasts on walls and floors as well.

If the light is distributed uniformly across room surfaces, glare can be avoided. The eye will adapt to a higher illuminance, usually producing an increase in visibility.

Strategies for avoiding glare include:

Controlling direct-daylight penetration

√   Although direct daylight can create dramatic changes in the environment, its intrusion in interiors should be limited to spaces that transition from outside to inside. Vestibules, lobbies and atria adapt well to direct daylight. Spaces where occupants can move freely to avoid direct sunbeams, or where typical occupancy is short or visual tasks are relaxed—such as malls, cafeterias, lounges, exercise rooms and corridors—also adapt well to direct daylight.

√   Control direct solar penetration with shading devices such as overhangs, fins and awnings located on the outside of the building, which allows daylight from the sky to be utilized with fewer thermal penalties.

√   Low-angle winter sun can be utilized as a passive solar heating strategy, but should only be considered in non-work spaces where glare would not be a problem performing critical visual tasks.

Controlling window glare

√   In addition to exterior sun control, interior window treatments, such as blinds and shades, should be used to block occasional direct daylight penetration and to modulate window brightness caused by bright and overcast skies.

√   The visual light transmission of the glass should be no higher than necessary to achieve daylighting goals.

√   Provide a comfortable gradient of brightness from the window to the interior by using splayed walls at the edges of windows.

√   If large areas of solid walls surround windows, consider lighting the wall with electric lighting to reduce contrast between the bright window and the wall that appears in silhouette. This strategy also applies to skylights and surrounding ceiling.

√   Use white or light-color matte or eggshell finishes on all surfaces near windows so that shiny surfaces do not become additional sources of glare.

Figure 14-6. *Electric lighting brightens the window wall at night. (Used by permission of Gruzen Samton Architects LLP. Project: US Environmental Protection Agency Building, Washington, DC. Lighting designer: Hayden McKay, Architects: Gruzen Samton Architects LLP; Croxton Collaborative)*

How to Optimize Daylighting for Different Building and Room Shapes

Daylighting is most successful, from a quality perspective, in shallow room configurations (no more than 8 meters, or 25 feet, from the windows) or private perimeter offices. This is because daylight can illuminate all of the room surfaces and raise the eye's adaptation level.

Most buildings constructed in North America before the 1940s were designed for daylighting. During the following era of cheap and plentiful energy, deep buildings began to flourish and daylighting declined.

In very deep spaces, the windows can become sources of glare in the distance without contributing to surface brightness near the viewer. Electric lighting must make up the difference, and at higher illuminances than if the space were windowless.

- √ High ceilings and placement of daylight windows at the top of the wall allow daylight to illuminate ceilings and room surfaces without being absorbed by furniture or partial-height panels.

- √ Continuous windows in horizontal bands can provide more visual comfort than "punched" windows.

- √ Adding windows to sidewalls or providing skylights can mitigate the high contrast between a window and the adjacent wall.

- √ Consider separating windows used for daylight and view to reduce thermal impacts and glare.

  - Windows located high on the wall are most suitable for daylighting and need a high visual light transmission capability (70-80%); they can also be fitted with light shelves and solar control devices.
  - Windows located lower on the wall are most suitable for views and can be specified with glass that has a lower visual light transmission capability (30-40%). Note that visual light transmission ratings significantly below 30%, particularly with smaller windows in regions with overcast climates, can be perceived as gloomy, thereby reducing the positive benefits of connecting to the out-of-doors.

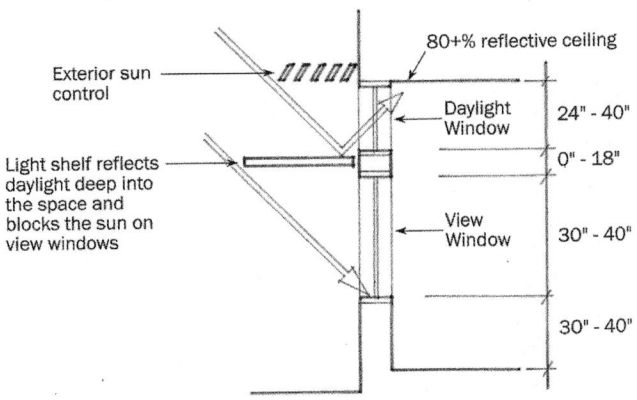

Figure 14-7. *Wall section for daylighting showing exterior shielding strategies.*

Figure 14-8. *Example of effective daylighting using a sawtooth roof. (photographer: Larry Lefever Photography)*

√ Consider supplemental toplighting strategies such as skylights, monitors or clerestories to deliver daylight deeper into windowed spaces or interior rooms.

- Skylights with wide-splayed wells spread light more evenly and reduce contrast of the bright skylight against surrounding surfaces. Splayed wells also help to minimize absorption of light within the well, increasing the efficiency of the daylight aperture and well. Diffuse skylights prevent direct daylight penetration and distribute light more evenly, although a view of the sky is lost.
- For any space, there is a percentage of ceiling space that can be dedicated to skylighting that optimizes the energy savings (typically 2-4%). Software is available to determine this percentage.
- Clear glass can be used in other forms of toplighting such as clerestory windows or monitors because it is easier to control direct daylight penetration through vertical than horizontal glass. Clear glass should not be used in clerestories where direct sunlight would cause a problem, however, and is best used in north-facing clerestories.

√ In historic buildings, where the windows are widely placed because of bearing wall construction, it may be necessary to wash the solid walls with electric lighting to reduce contrast and improve visual comfort.

Figure 14-9. *Example of a skylight. "Casa Nueva." (architect: Blackbird Architects; photography: © Bill Dewey)*

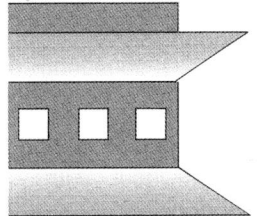
Punched windows create a sharp contrast between the window and surrounding wall.

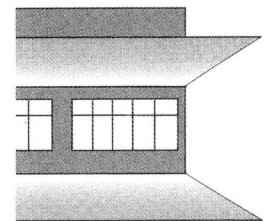
Continuous horizontal windows soften the contrast between the window and the wall.

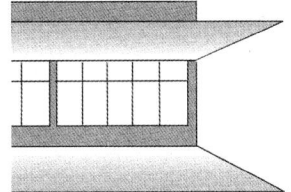
Window heads flush with the ceiling eliminate the wall above the window and soften the contrast between the ceiling and the window.

Figure 14-10. *Illustration of how different window configurations affect apparent window contrast. (photo from publication* Daylighting Design Smart and Simple *courtesy of EPRI)*

√ For daylight harvesting, *daylight zones* are defined as areas near windows where the electric lighting can be reduced or switched off when daylight is available. This is typically 4-5 meters, or 12-15 feet, from the window wall or a distance 1.5 times the window height.

How to Balance Electric Light with Daylight

When designing a luminous environment that balances daylight and electric light, it is recommended to consider daylight as the primary source of illumination. The electric lighting is designed to supplement the lighting when availability of daylight is limited, and to operate as stand-alone lighting at night.

√ Luminaires should provide light on the same surfaces and in a similar direction as the daylight apertures.

√ Diffuse sources, such as fluorescent, match the diffuse distribution characteristic of daylight well.

√ In deeper daylighted spaces, supplemental wall washing at the rear wall and adjacent ceiling softens the gradation of light from the window to the interior.

√ Because a positive daylight attribute is its continuous and full color spectrum, the color of the window glass should be as neutral as possible to avoid altering the transmitted spectrum.

√ The color temperature of electric light sources should match the color temperature of the daylight when applicable. The color temperature of daylight is very cool, ranging from 5000K to 10,000K at different times. When warm color temperature fluorescent sources (<3500K) are used in conjunction with very cool daylight (>5000K), they may appear yellow. The

Figure 14-11. *Metal halide luminaires accent the artwork on walls between the windows in this application. (Used by permission of Gruzen Samton Architects LLP. Project: US Environmental Protection Agency Building, Washington, DC. Lighting designer: Hayden McKay, Architects: Gruzen Samton Architects LLP; Croxton Collaborative)*

below recommendations are based on typical North American preferences and practice.

- If a space is occupied primarily during the day, then providing supplemental electric lighting (4000K-6000K) could be a reasonable choice for blending electric light with daylight.
- If the space is used in the evening or located in an overcast or cold climate with short winter days, however, most occupants find cool-colored lamps unpleasant; in such cases, neutral sources (3500K-4100K) may be favorably received in spaces that are daylighted.

How to Save Energy with Daylighting

√ Consider daylight harvesting control strategies such as switching or dimming. In a daylight harvesting control system, photosensors detect the contribution of daylight to a space. When light levels cross a target threshold, the control system responds by switching or dimming the lighting system in the designated control zones. For more information about lighting controls, see Chapter 10.

√ All complex control systems must be properly designed, installed and commissioned to provide sustained occupant acceptance and energy savings. For example, photosensors must be calibrated. It is recommended that these controls be recalibrated and commissioned periodically as part of their maintenance.

√ Avoid daylighting designs that can produce unacceptable lighting conditions, such as glaring windows, which may result in occupants taking action, such as closing blinds or shades, that can reduce energy savings potential.

√ Skylights and clerestories that eliminate direct sunlight can offer the most effective energy savings.

More Information

For more information, see references such as IESNA Daylight publications, EPRI's *Daylight Design—Smart and Simple* (available thru IESNA), "Tips for Daylighting" at http://btech.lbl.gov/pub/designguide, and Daylight *in Buildings: A Source Book on Daylighting Systems and Components*[79]).

LIGHT + CODE

If you have knowledge, let others light their candles by it. – Margaret Fuller

Code Compliance, Standards and Legal Issues

Energy codes are designed to set minimum energy performance standards for design and construction and can significantly reduce building system life-cycle costs.

Energy codes are considered an effective means of increasing overall building lighting efficiency because they typically establish minimum efficiency standards and control requirements and then allow designers to select the most appropriate products and technologies for the application within these boundaries. Some designers regard them as a necessary evil while others regard them as simply another set of requirements.

Building electrical codes are designed to establish minimum safety standards for design and construction to mitigate the possibility of fire and shock hazard during installation and use of electrical equipment.

This chapter describes relevant energy and building codes in the United States, Canada and Mexico.

United States

Energy codes

U.S. energy codes address lighting by setting lighting power density (LPD) limits on lighting and requiring mandatory lighting controls. Setting LPD limits for whole buildings is important because energy-efficient lighting can be inefficient as a whole if installed in high densities in a building. Energy codes are developed based on models that assume the use of available energy-efficient equipment, so the designer will have difficulty complying with codes if they specify inefficient equipment.

The most common national energy codes and standards are the ASHRAE/IESNA Standard 90.1 and the International Energy Conservation Code (IECC). The IECC applies to both residential and commercial buildings, while ASHRAE/IESNA Standard 90.1 covers only non-residential buildings. The U.S. Department of Energy (DOE) web site (www.doe.gov) offers publicly available software to help designers and engineers show compliance with the most common national codes and standards.

The ASHRAE/IESNA Standard 90.1 lighting energy standards, for example, were developed by considering design issues, illuminance requirements, tasks to be performed in the space and end-user needs. (For information about the methodology that was used to develop the lighting power densities for the ASHRAE/IESNA Standard 90.1,[80] visit the IESNA web site at www.iesna.org.) In addition to power density requirements, this Standard establishes mandatory requirements for the use of lighting controls and, in some cases, it limits the efficacy of allowable light sources (expressed as lumens per watt). There are requirements for both building interiors and exteriors.

Energy codes and standards vary significantly by state and even sometimes by municipality. The U.S. Department of Energy maintains a website at www.energycodes.gov, and the Building Codes Assistance Project has its own website at www.bcap-energy.org. These sites provide additional information on training, compliance, code adoption, implementation and technical support.

Building codes

In the United States, each state or municipality is responsible for creating or adopting its own building code. The National Fire Protection Association (NFPA) develops model codes that municipalities typically adopt with or without amendments.

NFPA 70, the National Electrical Code (NEC), sets minimum standards for the installation of lighting-related equipment and electrical safety as well as power for emergency egress lighting. NFPA 101, the Life Safety Code, defines allowable paths of egress, placement and visibility of exit signs, as well as minimum lighting requirements during egress conditions (see Chapter 16).

Many jurisdictions in North America require electrical equipment to be installed in accordance with an electrical code. Once installed, it is common practice for jurisdictions to require the installation to pass an electrical inspection before an occupancy permit is issued. While each locality may adopt its own unique electrical installation code, most are based on the NEC in the U.S., the Canadian Electrical Code, Part 1 in Canada, or the Mexican Electrical Code in Mexico. To aid these jurisdictions in performing electrical inspections on installations that contain a multitude of different types of lighting products, electrical codes have established a process for listing, certifying or approving individual electrical products in accordance with an established safety standard. In the U.S., Underwriters Laboratories Inc. (UL) is the organization that is accredited by the American National Standards Institute (ANSI) to establish and maintain safety standards for lighting products. In Canada and Mexico, the respective organizations responsible for developing safety standards are the Canadian Standards Association (CSA) and the Associón de Normalización y Certificación AC (ANCE).

UL, CSA and ANCE develop standards for the investigation of products in relation to hazards to life and property with a focus on minimizing a product's risk of fire and risk of shock. While using a product that is listed, certified or approved by UL, CSA, ANCE (or one of the other organizations that is accepted to local jurisdiction) does not assure a safe electrical installation, the use of products that bear an approval mark from an organization that tests in accordance with the applicable UL, CSA or ANCE standards does indicate that the product itself is not inherently dangerous.

CHAPTER 15: LIGHT + CODE

The Energy Policy Act of 2005 (EPAct 2005)

EPAct 2005 includes several important lighting-related provisions, including an accelerated tax deduction for commercial buildings as an incentive to meet certain efficiency targets, funding for research and testing of advanced buildings, energy efficiency standards for select lighting products, and efficiency standards that apply to Federal buildings.

Americans with Disabilities Act (ADA)

The Americans with Disabilities Act (ADA) provides guidelines for making buildings safer and more accessible for disabled persons. The ADA requires, for example, that wall sconces mounted below 2 meters (80 inches) above the floor not project more than 10 centimeters (4 inches) from the wall so that visually impaired or disabled people will not injure themselves by accidentally walking into them. ADA guidelines are also currently being incorporated into the International Building Code (IBC), which is commonly adopted by individual localities as part of their building codes.

Canada

Energy codes

The Model National Energy Code of Canada for Buildings (MNECB 1997) is prepared by the Canadian Commission on Building and Fire Codes (CCBFC) and is published by the National Research Council (NRC). Under the Constitution Act, regulation of building in Canada is the responsibility of provincial and territorial governments; MNECB, therefore, is prepared in the form of a model code to permit adoption by the appropriate authority.

The MNECB differs from model codes traditionally produced by CCBFC in that it addresses environmental protection and resource conservation issues rather than the health and safety of occupants. The MNECB essentially establishes minimum requirements for energy efficiency in buildings and establishes a standard of construction for those features of buildings that affect their energy efficiency. Energy efficiency in small residential buildings is addressed by a companion document, the Model National Energy Code of Canada for Houses (MNECH).

The MNECB defines requirements for the design and construction or specification of lighting in Part 4, which includes lighting in interior spaces, building exteriors, exterior building areas such as entrances, exits, and loading docks, as well as grounds, parking and other exterior areas associated with the building site. The MNECB also addresses specific situations such as controls for hotel guest rooms and lighting for special spaces and activities such as rooms serving multiple functions, containing multiple simultaneous activities and containing activities of indoor sports.

Part 4 of the MNECH, which applies to small residential buildings, also contains requirements intended to minimize energy use for lighting and to encourage the provision of means of efficiently controlling lighting. The type of lighting addressed are similar to the MNECB except that lighting for grounds, parking and other exterior areas apply only to common exterior areas of multiple-unit residential buildings or building complexes.

Building codes

The MNECH and the MNECB are intended to be used in conjunction with the National Building Code (NBC), which is a code of minimum regulations for public health, fire safety and structural sufficiency.

Lighting requirements in the 1995 National Building Code:

- The National Building Code of Canada (NBC) is a set of minimum provisions for the safety of buildings. It is applicable to the construction of new buildings and the renovations of existing buildings. The NBC is a model code that is adopted by the provinces and territories with or without amendments.
- In the NBC, buildings may fall under either Part 3 or Part 9 depending on their size and use. The size criteria for a part 3 building include buildings more than three stories in height or more than 600 square meters in area on largest floor. Also, assembly-type buildings (arenas, stadiums, performing arts centers), institutional occupancies (jails, hospitals) and hazardous industrial occupancies are to be constructed to the provisions of Part 3 regardless of their size. Generally, small residential buildings fall under Part 9 of the NBC.
- Parts 3 and 9 of the NBC contain very minimal requirements on lighting (50 lx minimum required along egress routes and 10 lx minimum required where emergency lighting is required). None of these requirements are in relation to the saving of energy as it is not one of the objectives of this code.
- Section 3.8 of the NBC addresses the minimum requirements on barrier-free design. It does not contain requirements on lighting. It should also be noted that not all provinces and territories on Canada adopt Section 3.8, as some have developed their own requirements.

Mexico

Energy codes

Energy codes in Mexico are divided basically in two categories: federal codes and local codes, the federal being the most important. Federal codes were introduced in the 1990s and have been revised continuously since then. They address energy efficiency only, and are not intended to set guidelines for lighting design. Local codes are usually intended for general building purposes and include a section on energy efficiency.

Lighting professionals in Mexico usually refer to IESNA standards for design criteria and recommended illuminance values. In considering local needs, however, some revised IESNA tables have been introduced by Mexican professionals. For example, recommended street lighting illuminances in Mexico are usually higher than those used in the U.S. for several reasons—safety, larger numbers of pedestrians, and ambience.

In 1990, the Energy Saving Fund (FIDE) was created to support actions that reduce energy consumption. In general, FIDE works by supporting projects in which energy savings will provide a sufficient return on investment. The fund covers part of the initial cost and the equivalent to savings is paid by users in their energy bill until the investment is paid off.

Federal codes can be reviewed at the Energy Ministry website, www.sener.gob.mx, or at the National Codes website, www.economia-nomx.gob.mx. Federal codes are known as NOM (Norma Oficial Mexicana) and they include a number, the issuer entity, the year of introduction and its name. The NOMs related to lighting include:

- NOM-007-ENER-2004, Energy Efficiency for Lighting Systems in Non-Residential Buildings

  The most important reference in Mexican codes for lighting system energy considerations, this code's purpose is to "reduce energy consumption and contribute to preservation of energy sources as well as the Nation's ecology." It applies to office buildings, schools, commercial buildings, hospitals, hotels, restaurants, warehouses, recreational facilities, workshops, transport facilities, etc. Buildings not covered include show business installations, cinema and TV sets, exhibition halls, and other applications with special needs. It includes a glossary with vocabulary related to lighting and building classification, a figure with power density recommendations, some recommendations for facade lighting and parking lot lighting. It establishes the calculation method for power density.

- NOM-017-ENER-1997 Energy Efficiency in CFLs, Limits and Testing Methods

  This code, presented by the Energy Secretary, addresses the minimum standards for compact fluorescent lamps and related ballasts to be used in Mexico. Lamps less than 28 watts are covered.

- NOM-013-ENER-2004 Energy Efficiency for Outdoor and Street Lighting

  This 2004 code from the Energy Secretary primarily limits the maximum power density for exterior lighting at public facilities such as streets, roads, parking areas (open and closed), gardens, parks, etc. It excludes airports, emergency lighting, seasonal lighting, amusement parks, sea platforms, temporary building installations, sign boards, docks and tunnels. Its glossary defines vocabulary for lighting and outdoor spaces. The first figure establishes maximum power densities and recommendations for light levels relative to street width. The second figure modifies these values for places where poles higher than 18 meters (54 feet) are used. The code includes a calculation method for obtaining power density values.

Building codes

Released in 1999 by the Labor Secretary, NOM-025-STPS-1999 Lighting Conditions at Working Centers was implemented to prevent workers at labor centers from having any health risks produced by a deficient lighting system. It includes a glossary that defines the vocabulary involved both in lighting and in labor activities. It sets the responsibility for business owners to register, evaluate and control workplace lighting levels, and establishes their obligation to inform workers about the risks that glare and insufficient lighting may produce. Business owners are also obligated to create a luminaire maintenance program as well as implement an emergency lighting system, especially in those places where darkness due to a blackout could be hazardous. There is a table of recommended illuminances for different visual tasks and spaces. Finally, the code sets limits to reflected glare and direct glare, and outlines a method to evaluate lighting levels in the field.

16

LIGHT + SAFETY

Freedom is just Chaos, with better lighting. – Alan Dean Foster

Safety, Security and Emergency Egress

Safety and security for occupants and visitors should be a top priority in all lighting designs. Three levels of consideration are safety from bodily injury, safety from hurt or loss, and safety during emergency egress. Local codes may pose requirements.

This chapter describes lighting for safety and security and provides recommendations for achieving effective solutions.

Safety from Bodily Injury

Features in potentially hazardous areas—such as stairs, bathtubs and shower stalls, train platforms, crosswalks across streets—and industrial areas where machinery or potentially dangerous manufacturing processes occur—should be clearly visible.

How to Light for Safety

√   In hazardous areas, such as stairs or around dangerous equipment, use a combination of high-contrast markings and light patterns that reinforce these markings.

Figure 16-1. *The lighting on these stairs produces a pattern of light and shadow that clearly differentiates the riser and tread. (lighting design: exterior-Janet Lennox Moyer, interior-Naomi Miller; photographer: Kenneth Rice Photography)*

Safety from Hurt and Loss

Lighting can have a very positive influence on personal security. Lighting alone, however, does not deter crime or prevent accidents (if it did, there would be no crime in the daytime), nor is lighting appropriate everywhere in all applications.[81, 82] Current evidence suggests that lighting may not eliminate crime, but rather change the type of crime committed or move it to another location.[83]

Lighting can provide three positive effects related to incidence of crime and personal security. First, it can increase pedestrian activity. Second, it can make the criminal feel conspicuous and vulnerable to identification. Third, it can make the potential victim feel safer because he or she is able to identify danger more quickly and react to avoid harm.

How to Light for Security

√ *Provide maximum, but practical, illuminances on the street, parking lot or walkway surface.* 30 lux (3 fc) may be considered a point at which people feel comfortable in urban areas, irrespective of how the light is delivered; illuminances higher than 30 lx (3 fc) do not significantly increase comfort.[84] This illuminance, however, is unnecessarily high for nearly all outdoor visual tasks, and if all outdoor spaces were lighted to 30 lx (3 fc), there would be a high cost in terms of electrical energy and additional light pollution. In addition, with lower ambient levels, less light is needed for most visual tasks. Experienced designers can deliver a perception of safety at lower illuminances, but how and where the light is applied becomes critical.

√ *Provide uniformity of illuminance.* An average-to-minimum ratio of 4:1 or less appears to produce higher perceptions of safety. Uniformity is also strongly related to vertical illuminance and the ability to see and recognize faces.

Figure 16-2. *(left)* **Figure 16-3.** *(right) Two parking lots with similar average illuminances. One has better uniformity than the other. Which one would make you feel more comfortable because you could see somebody standing anywhere within a 16-meter (50-foot) range? (Photos courtesy of Nancy Clanton)*

Figure 16-4. *(left)* **Figure 16-5.** *(right) In which garage would you feel safer? Why? (photographer: Naomi Miller)*

- √ *Provide perimeter lighting.* Light the perimeter of an area so that it is easy for a pedestrian to identify potential danger and know where to flee for refuge. Lighted walls and surfaces can help increase the perception of safety.

- √ *Eliminate glare.* Choose luminaires with distributions that limit luminous intensity (candlepower) at the angles most likely to produce the sensation of glare, or mount the luminaire such that offending angles of light are blocked from view. In most cases, the glare angles for a parking lot or streetlighting luminaire are between 80° and 90°. Eliminating glare in the field of view can result in a dramatic improvement in vision at lower illuminances. For more information, see IESNA RP-33-99 and IESNA RP-8.[85, 86]

Safety during Emergency Egress

A building may be evacuated during an emergency. In some emergencies, exiting the building is a visual task that is matter of life or death. Occupants and visitors, however, may become panicked and not think clearly, or building areas may fill with smoke, making egress difficult. This is where emergency lighting provides a critical service.

Emergency lighting systems indicate the path of egress and illuminate escape routes. Properly designed, these systems direct occupants safely to exits. Conversely, emergency lighting that is not coordinated with the building layout can slow down or prevent rapid evacuation of a building.[87]

How to Light for Emergency Egress

- √ Comply with local life safety building codes. This is only a minimum level of design, however. Envision a panicked occupant in a building with the lights on, the lights off, the building filling with smoke. How easy is it to find and navigate the egress path?

- √ Ensure the emergency lighting does not create glare situations or other impairment to visibility.

- √ Be aware of occupants with special needs, such as vision- or hearing-impaired people, who may need additional notification systems.

Figure 16-6. *Various exit signs demonstrate different levels of visibility in smoky conditions. (Photo courtesy of Lighting Research Center.)*

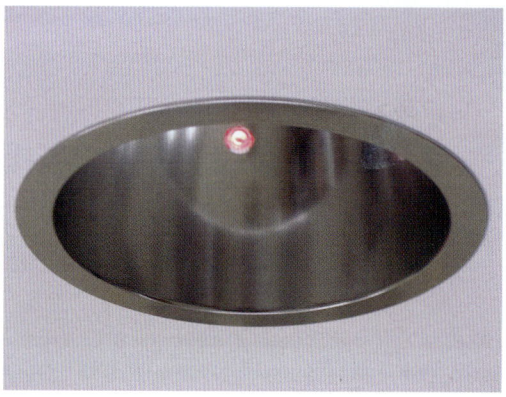

Figure 16-7. *Downlight with test button for emergency option.*

√ Specify exit signs that are highly visible in different lighting and smoke conditions.[88, 89]

√ Specify emergency lighting equipment that is easy to maintain and test. Often, emergency systems are only tested when an emergency occurs. Some products are self-testing to comply with many life safety and other applicable codes. Some products are also self-diagnostic, providing visual information about the unit's status.

√ Emergency lighting is often criticized as an unattractive addition to the lighting design. However, emergency lighting can be integrated into architectural lighting systems by 1) adding emergency ballasts to architectural lighting systems, 2) specifying emergency luminaires that offer the same appearance as conventional downlights, 3) connecting a small number of luminaires to a power supply that will switch to emergency power during a primary power outage, and 4) specifying exit signs that present sleeker shapes and more compatible finishes.

SIDEBAR 16-1. Emergency Lighting in the World Trade Center in 1993

In 1993, a bomb placed in a truck detonated in the underground garage of the World Trade Center. The blast destroyed the emergency generator system as well as the power supply to the two 110-story towers. It took up to five hours to evacuate more than 20,000 people working in each tower.

Why did it take so long?

To evacuate the building, occupants had to navigate 100 floors of darkened, crowded stairwells. The power to the emergency lighting in the three stairwells of each tower vanished when the generator was destroyed.

The New York Port Authority, responsible for operating the buildings, repaired the damage with a distributed emergency lighting system. Instead of one power supply for the emergency lighting in the entire building, independent battery-powered emergency lights in the stairwells were used. A single blast could now only destroy the lighting at one floor, not the entire stairwell. The Port Authority also installed photoluminescent paint and signs that have the property to store light and glow in the dark in the event of a power failure.

During the terrorist attack of September 11, 2001, occupants were able to evacuate the building efficiently; nearly 17,000 people evacuated the buildings before they collapsed.

If not for the lessons learned from the 1993 bombing, the loss of life in 2001 would have been even more devastating.

LIGHT + SPECIAL CONSIDERATIONS

To be a star, you must shine your own light.... — Unknown

Special Considerations

Any project may encounter a number of important lighting issues that are unique to the project. This chapter describes a few examples of these special considerations, including lighting precious artwork and artifacts, lighting for cameras, and addressing luminaire noise:

Conservation of materials (museums)

Visible light and ultraviolet and infrared energy can damage organic materials over time. In museums or galleries that display objects such as important paintings, textiles or artwork, it is important to limit daylight and electric light exposure. This is done by reducing illuminance on the object, reducing the amount of time the object is exposed to the light, and minimizing ultraviolet and infrared radiation (which do not contribute to visibility) on the object through filtering or reflection.

See IESNA RP-30-96, *Recommended Practice for Museum and Art Gallery Lighting*, for more information.[90]

Lighting for cameras

Security cameras are used in lobbies, corridors, parking garages and outside of bank and retail establishments to monitor occupancy. Lighting must be provided to ensure that there is sufficient illuminance on faces and bodies of occupants and visitors so that security cameras can record a scene and enable identification.

Camera needs for lighting may also supersede human needs for light in a television or film studio. In these applications, it is important to ask what vertical or horizontal illuminances are needed, where they are needed, and whether the color spectrum of light is important.

Luminaire noise

Luminaires, specifically the components they house such as ballasts or transformers, can produce an audible hum or buzz.

This is often a negligible issue, although if the space is acoustically sensitive, such as a concert hall or music library, it can be distracting and interfere with the music.

In sporting arenas, pools or courts, especially where there are few sound-absorbing surfaces to attenuate the hum from lights, consider ballasts with the best sound ratings. This may mean specifying ballasts that have extra "potting" in them, resilient pads or mountings so that the ballast does not telegraph vibration to other parts of the fixture or structure, or electronic ballasts. In many applications, the ballast can be mounted remotely in a closet or attic. (Note that older sound ratings—A, B, C, etc.—for magnetic ballasts are not applicable to electronic ballasts.)

In concert halls and auditoriums, follow the same advice, but it may be necessary to consider the interaction of the dimming system with the lighting components. Properties of dimmers, such as rise time and the power sine wave, can affect noise, even with incandescent lamps that do not use ballasts and transformers. Dimmer rack fans can also generate noise.

Fans from illuminators used in fiber-optic installations can also be a source of noise. Keep these illuminators acoustically isolated or use fan-less versions.

PART FOUR: LIGHT + APPLICATION

*There are two ways of spreading light—to be the
candle or the mirror that reflects it.*

– Edith Wharton

A NEW PROJECT: COLLECTING INFORMATION

Client Goals and User Needs

The first step in any design is to gather information about user needs and the client's goals. Without this information, the lighting solution for a retail store selling designer sneakers would be no different from that of an iron foundry or a kindergarten classroom.

This chapter provides a preliminary list of questions the designer should ask the client, other design team members and him/herself. The answers will help to identify the most important design issues. Note, however, that this is by no means a complete list.

The WHO of Lighting Design

- √ Who will be judging the project's success, and what are their aesthetic preferences, expectations and priorities (i.e., cost, energy, ease of maintenance, appearance, productivity, safety, etc.)?

- √ Who will be using the space? What are their ages and visual capabilities? What visual "work" are they performing, and what are their stylistic tastes and expectations? Remember that "users" may be both staff and customers, performers and audience, supervisors and workers, health care workers and patients, night security and cleaning personnel, and so on.

- √ Where are the users located, and where will they be looking? For example, it may be wise to avoid recessed downlights in a hospital corridor because of the glare they may produce for patients lying on a gurney.

Human Needs

- √ What visual "work" are these users performing? How should the task lighting and/or general lighting be designed to improve visibility? Where are the visual tasks performed, and can difficult tasks be illuminated with localized lighting rather than general lighting?

√ Is visual comfort an issue for the users? How can the lighted environment prevent annoying or disabling glare?

√ What kind of atmosphere is being created? Should it appear bright and busy, dim and relaxed, cheerful, safe, spooky, romantic, etc.? How can room surface brightness be used to affect this perception? Are there any other psychological issues that need to be addressed with good lighting?

√ Will sparkle enhance the appearance of a space or displayed objects, or can reflected highlights help a user see important details in an industrial task?

√ Will shadows on the task, faces, objects or building materials be a problem or an enhancement? What kinds of luminaires will produce helpful shadows and eliminate confusing shadows?

√ Does flexibility of activities, visual tasks, task locations and/or mood suggest the use of separate lighting control circuits or a variable dimming system?

√ Does illumination need to be uniform on the task plane to accommodate flexible task locations, or is some variation acceptable?

√ Is face or body appearance important to communication, feelings of well-being or sales? (Example: How does the customer look in the mirror of the fitting room in a clothing store?) Does the lighting system create a pleasant modeling of light for the user or viewer?

√ Is color appearance of the visual task, space and/or people in the space important?

√ Is flicker or strobe likely to cause a health, safety and/or annoyance problem to users?

√ Is good peripheral vision a safety issue for users or security patrols?

Economics, Energy and the Environment

√ Is project cost an issue? What are the budgets for lighting, and are they appropriate for the type of project? How much flexibility does the budget have for adjustment?

√ Is operating cost an issue? If so, how can lighting maintenance and energy use be kept at a reasonable level?

√ What are the applicable energy codes for lighting? What is the local cost of lighting energy? Can more efficient lighting products or controls be implemented to reduce energy use?

√ What are the applicable life safety codes regarding emergency egress? This sets a minimum standard for design. How can the lighting work with the environmental design to improve safety and personal se-

CHAPTER 18: A NEW PROJECT: COLLECTING INFORMATION

curity for users? How can the lighting help maintain security of the physical property?

√ Is the project required to achieve certain sustainability goals defined within a green building rating system such as LEED?

√ Is light spilling onto neighboring properties a potential problem? Will light spill from this project interfere with astronomical observations or appreciation of night skies?

√ Are there lighting products that are more environmentally friendly choices? They may be locally produced, built from sustainable materials, particularly energy effective, easily separated into bins for recycling at the end of useful life, or they may have less toxic impact during manufacture, use or disposal.

√ Can more be done with less? In areas with low ambient illuminances, a little light can go a long way. In areas with high ambient illuminances, do not just keep raising the ante—consider alternate means of attracting notice such as selective highlighting, color or movement.

Architecture and Other Building-Related Issues

√ What are the space dimensions? Where will departments, furniture and equipment, and different types of visual tasks be located?

√ What are the important visual elements or features in the space, and how can the lighting products and patterns of light support these? These elements can include art objects or important signage, or even simple room numbers. (At the very least, the lighting system and its lighting patterns should not distract from important visual information in the space.)

√ What are the room surface finishes in terms of color, reflectance and specularity?

√ What is the style of the space? Is there a planned décor that the lighting should support or highlight?

√ If the space already exists, what lighting is currently installed? Can some of it be reused? Do users have any comments on the existing lighting that can help guide improvements and changes?

√ Will this space be used during the day or night? How can daylight be admitted through windows or skylights without excessive glare or heat? Can daylight and electric light be coordinated for best appearance, function and energy savings?

√ Do any operating conditions warrant special consideration in the lighting design, such as hazardous conditions or the area being prone to vandalism?

√ How will the lighting be maintained? Do any special maintenance requirements exist?

19

LIGHTING DESIGN GUIDE

Lighting Design Guide

Older editions of the IESNA *Lighting Handbook* recommended illuminances for specific applications and visual tasks, and as a result many designers often perceived the IESNA system of recommended illuminances as the primary, or even sole, crite¬rion for an effective lighting design.

The Ninth Edition of the *Lighting Handbook* introduced the IESNA Lighting Design Guide (2000)—a new system for considering a wide range of lighting design criteria that tran¬scend illuminance recommendations and address a range of critical lighting factors.[91] Properly followed, these criteria in¬crease the quality of the visual environment.

With the publication of LIGHT + DESIGN, the 2000 Lighting Design Guide tables have been superseded by new Lighting Design Guide tables, which were developed to update IES lighting recommendations based on evolving thought in design. For example, the 2000 Lighting Design Guide tables do not address energy efficiency, environmental considerations, cost effectiveness and maintenance; the updated Lighting Design Guide addresses these issues for certain applications/tasks.

While the 2008 Lighting Design Guide offers guidelines, LIGHT + DESIGN is intended to support its utilization by helping designers become educated about the issues involved and adjust priorities according to the project needs.

Factors that might lead to deviations from the recommendations include life/safety, security and health issues (including lighting for the aged and visually impaired); energy requirements; historical context; unusual maintenance requirements; lighting for plants, animals and cameras; and unusual client specifications. The designer is strongly encouraged to document and explain all deviations from the recommendations in the Lighting Design Guide.

In addition, the reader is encouraged to consult other IES publications, such as the series of Recommended Practices, for additional information, and the *Lighting Handbook* for reference and greater understanding of recommended illuminances.

This chapter introduces the Lighting Design Guide with guidelines for its proper use.

How to Use the Lighting Design Guide

The Lighting Design Guide is divided into six sections:

I. Interior
II. Industrial
III. Outdoor
IV. Sports and Recreation
V. Transportation
VI. Emergency, Safety and Security

Within each section, there are headings, subheadings, and *location* and *task* listed alphabetically and presented in rows. For example, in the Interior section, Educational Facilities is listed, and under this heading is a subheading titled Classrooms, under which is a location termed Science Laboratories.

The columns present criteria important for a high-quality visual environment, including both horizontal and vertical illuminances. One of four levels of importance is entered in the cells of the matrix: Very Important, Important, Somewhat Important and Not Important/Not Applicable. Each cell entry represents the relative importance of one criterion for a particular *location* or *task*.

For example, consider the lighting of an open plan office. The designer would reference the Interior section of the Guide, find the Offices heading, then the Open Plan Office subheading. The designer will learn that if the office features intensive use of computer screens, vertical illuminance, source/task/eye geometry, reflected glare, visual comfort and room surface brightness are considered Very Important design factors.

CHAPTER 19: LIGHTING DESIGN GUIDE

Summary of Differences between 2000 and Updated Lighting Design Guide

2000 Lighting Design Guide	Updated Lighting Design Guide	LIGHT + DESIGN Reference
	Lighting for Human Needs	
Source/Task/Eye Geometry Reflected Glare Intrinsic Material Considerations Shadows	Task Visibility 　Source/Task/Eye Geometry 　Reflected Glare 　Disability Glare 　Shadows 　Intrinsic Material Considerations	Chapter 2: Light + Vision
Direct Glare	Visual Comfort 　Discomfort Glare 　(including Overhead Glare)	Chapter 3: Light + Visual Comfort
Flicker and Strobe	Flicker and Strobe	Chapter 3: Light + Visual Comfort
Light Distribution on Task Plane (Uniformity)	Light Distribution on Task Plane (Uniformity)	Chapter 13: Light + Distribution
Modeling of Faces and Objects	Modeling of Faces and Objects	Chapter 4: Light + Modeling
Color Appearance (and Color Contrast)	Color Appearance (of Objects or People or Light Sources)	Chapter 5: Light + Color
Peripheral Detection	Peripheral Detection	Chapter 11: Light + Night
Sparkle/Desirable Reflected Highlights	Sparkle (or Desirable Reflected Highlights)	Chapter 13: Light + Distribution
System Control and Flexibility	System Control and Flexibility	Chapter 10: Light + Control
Special Considerations	Special Considerations	Chapter 17: Light + Special Considerations
	Lighting for Economics and the Environment	
	Cost, both Initial and Maintained	Chapter 6: Light + Cost
	Maintenance and Change	Chapter 7: Light + Maintenance
	Energy Use	Chapter 8: Light + Energy
	Environmental Considerations	Chapter 9: Light + Environment
	Controls for Energy	Chapter 10: Light + Control
Light Pollution/Trespass	Light Pollution/Light Trespass	Chapter 11: Light + Night
System Control and Flexibility	System Control and Flexibility	Chapter 10: Light + Control
Peripheral Detection	Peripheral Detection	Chapter 11: Light + Night
	Lighting for Architecture and other Building-Related Issues	
	Code Compliance, Standards, and Legal issues	Chapter 15: Light + Code
	USA	Chapter 15: Light + Code
	Canada	Chapter 15: Light + Code
	Mexico	Chapter 15: Light + Code
	Safety, Security and Emergency Egress	Chapter 16: Light + Safety
Appearance of Space and Luminaires	Appearance of Space and Luminaires	Chapter 12: Light + Architecture
Daylighting Integration and Control	Daylight and View	Chapter 14: Light + Daylight
Luminances of Room Surfaces	Room Surface Brightness (and Surface Characteristics)	Chapter 13: Light + Distribution
Light Distribution on Surfaces	Light Distribution on Surfaces (Patterns)	Chapter 13: Light + Distribution
Points of Interest	Points of Interest	Chapter 13: Light + Distribution
Surface Characteristics	Room Surface Brightness (and Surface Characteristics)	Chapter 13: Light + Distribution
Special Considerations	Special Considerations	Chapter 17: Light + Special Considerations
Light Distribution on Task Plane (Uniformity)	Light Distribution on Task Plane (Uniformity)	Chapter 13: Light + Distribution

IES Lighting Design Guide

Table 19-1. Updated Lighting Design Guide lighting recommendations for selection of visual tasks in Interior environments.

Legend: ■ Very Important | □ Important | ⊡ Somewhat important | Blank = Not important or not applicable

I. INTERIOR LOCATIONS AND TASKS

Location/Task	|	Lighting for Human Needs (Design Issues)												Lighting for Economics and the Environment						Lighting for Architecture and other Building-Related Issues							Illuminance (Horizontal)	Category or Value (lux)[a]	Illuminance (Vertical)	Category or Value (lux)[a]	Notes on Illuminance	Ref. Ch. in IESNA Lighting Handbook
		Task Vis. – Source/Task/Eye Geometry	Task Vis. – Reflected Glare	Task Vis. – Disability Glare	Task Vis. – Shadows	Visual Comfort – Discomfort Glare (incl Overhead Glare)	Flicker and Strobe	Light Distribution on Task Plane (Uniformity)	Modeling of Faces and Objects	Color Appearance (of Objects or People or Light Sources)	Sparkle (or Desirable Reflected Highlights)	Special Considerations	Notes on Special Considerations	Cost, both Initial and Maintained	Maintenance and Change	Energy Use	Environmental Considerations	Controls for Energy	System Control and Flexibility	Code Compliance, Standards, and Legal Issues	Safety, Security and Emergency Egress	Appearance of Space and Luminaires	Daylight and View	Room Surface Brightness (and Surface Characteristics)	Light Distribution on Surfaces (Patterns)	Points of Interest						
Offices (13)																																
Filing (see Reading)		□	■	■	□	■	□	□	□	□				□	□	□	□	□		□	□				□		□	E	■	C		Ch. 11
General and private offices (see Reading)																																
Open plan office		□	■	■	□	■	□	□	□	□		□	14,15	□	□	□	□	□		□	□	□	□	■	□		□	D	■	B		
Intensive computer use		□	■	■	□	■	□	□	□	□		□	14,15	□	□	□	□	□		□	□	□	□	■	□		□	E	□	B		
Intermittent (normal) computer use		■	■	■	□	■	□	□	□	□				□	□	□	□	□		□	□	□	□	■	□		□	E	□	B		
Private office						□																■	□	□	□		□	C	■	A		
Libraries (see Libraries)																																
Lobbies, lounges, and reception areas																																

Notes:

(a) Low illuminances (less than 30 lux) are given in lux; values greater than 30 lux are given in letter categories.

(1) Consider lighting for video cameras.
(2) Maximum illuminance
(3) At ground level
(4) The minimum illuminance in a prison cell is 200 lx (20 fc); 300 lx (30 fc) should be provided for reading at the head of the bed.
(5) Adjustable task lighting recommended.
(6) Task lighting recommended, possibly located in ceiling.
(7) Degradation factors important to consider.
(8) In the display plane.
(9) Special lighting for signage or banners may be required.
(10) See Chapter for special considerations.
(11) Refer to Chapter 16, Health Care Facilities Lighting for specific recommendations for surgical task lighting
(12) at 30" above floor
(13) Design issues, including illuminances, can be listed for room or space. Refer to specific task under Reading or Graphic Design and Materials, for example.
(14) Lighting should be flexible to accommodate changes in office furniture
(15) Acoustical aspects of luminaires need to be considered
(16) Design issues and illuminances listed for the task. See "Offices" or "Educational Facilities", for example, for additional considerations for the room.

Table 19-2. Updated Lighting Design Guide lighting recommendations for selection of visual tasks in Industrial environments.

II. INDUSTRIAL LOCATIONS AND TASKS[a]

Legend: ■ Very Important   ☐ Important / Somewhat important   Blank = Not important or not applicable

Design Issues	Assembly – Simple	Assembly – Difficult	Assembly – Exacting	Warehousing – Inactive	Warehousing – Active: bulky items; large labels	Warehousing – Active: small items; small labels	Service spaces – Stairways, corridors
Lighting for Human Needs							
Task Visibility – Source/Task/Eye Geometry	■	■	■	■	■	■	■
Task Visibility – Reflected Glare	■	■	■	☐	☐	■	☐
Task Visibility – Disability Glare	☐	■	■		☐	☐	■
Task Visibility – Shadows	■	■	☐	☐	☐		■
Task Visibility – Intrinsic Material Considerations							
Visual Comfort – Discomfort Glare (incl Overhead Glare)	☐	☐	■	☐	☐	■	☐
Flicker and Strobe							
Light Distribution on Task Plane (Uniformity)	☐	☐	☐	☐	☐	☐	☐
Modeling of Faces and Objects	☐	☐	☐	☐	☐	☐	
Color Appearance (of Objects or People or Light Sources)	☐	☐	☐	■	■	■	☐
Special Considerations							
Notes on Special Considerations							
Lighting for Economics and the Environment							
Cost, both Initial and Maintained	☐	☐	☐	☐	☐	☐	
Maintenance and Change	☐	☐	☐	☐	☐	☐	
Energy Use	☐	☐	☐	☐	☐	☐	
Environmental Considerations	☐	☐	☐	☐	☐	☐	
Controls for Energy				☐	☐	☐	
System Control and Flexibility							
Lighting for Architecture and other Building-Related Issues							
Code Compliance, Standards, and Legal Issues	☐	☐	☐	☐	☐	☐	☐
Safety, Security and Emergency Egress	☐	☐	☐	☐	☐	☐	
Appearance of Space and Luminaires	☐	☐	☐	☐	☐	☐	
Daylight and View	☐	☐	☐	☐	☐	☐	
Room Surface Brightness (and Surface Characteristics)	■	■	■	☐	☐	☐	
Light Distribution on Surfaces (Patterns)	☐	☐	☐	☐	☐	☐	☐
Illuminance Categories (See Note b)							
Illuminance on Task Plane[b]	■	■	■	☐	☐	☐	☐
Category or Value (lux)[c]	D	F	G	B	C	D	B
Notes on Illuminance – see end of section							
Ref. Reference Chapter(s) in IESNA Lighting Handbook	Ch.7	Ch.7	Ch.7	Ch.7	Ch.7	Ch.7	Ch.7

Basic Industrial Tasks

Notes:
(a) For details on specific tasks or spaces refer to Chapter 19, Industrial Lighting.
(b) The task may be horizontal, inclined, or vertical.
(c) Low illuminances (less than 30 lux) are given in lux; values greater than 30 lux are given in letter categories.

Table 19-3. Updated Lighting Design Guide lighting recommendations for selection of visual tasks in Outdoor environments.

III. OUTDOOR LOCATIONS AND TASKS

Legend: ■ Very Important   ☐ Important   ◻ Somewhat important   Blank = Not important or not applicable

Design Issues	Bikeways — Alongside roadways — commercial areas	Bikeways — Distant from roadways	Parks, Plazas, and Pedestrian Malls	Parking Areas
Lighting for Human Needs				
Task Visibility - Source/Task/Eye Geometry	■	■	■	■
Task Visibility - Reflected Glare	■	■	■	■
Task Visibility - Disability Glare	■	■	■	■
Task Visibility - Shadows	■	■	■	■
Visual Comfort - Discomfort Glare (incl Overhead Glare)	■	■	■	■
Light Distribution on Task Plane (Uniformity)			☐	
Modeling of Faces and Objects	■	■	■	
Color Appearance (of Objects or People or Light Sources)	■	■	☐	
Peripheral Detection	■	■	■	■
Special Considerations				
Notes on Special Considerations				
Lighting for Economics and the Environment				
Cost, both Initial and Maintained	☐	☐	☐	☐
Maintenance and Change	☐	☐	☐	☐
Energy Use	☐	☐	☐	☐
Environmental Considerations	☐	☐	☐	☐
Controls for Energy	☐	☐	☐	☐
Light Pollution/Light Trespass	■	■	■	■
System Control and Flexibility				
Lighting for Architecture and other Building-Related Issues				
Code Compliance, Standards, and Legal Issues	☐	☐	☐	☐
Safety, Security and Emergency Egress	☐	☐	☐	☐
Appearance of Space and Luminaires	☐	☐	☐	
Room Surface Brightness (and Surface Characteristics)	◻	◻	◻	
Light Distribution on Surfaces (Patterns)	☐	☐	☐	☐
Points of Interest	☐	☐	☐	◻
Illuminance Categories (Horizontal and Vertical)				
Illuminance (Horizontal)	◻	◻	◻	◻
Category or Value (lux)[a]			B	
Illuminance (Vertical)	■	■	■	■
Category or Value (lux)[a]			A	
Notes on Illuminance - see end of section	(1,8)	(1,8)		(4)
Ref. Reference Chapter(s) in IESNA Lighting Handbook	Ch. 21, 22, 29	Ch. 22	Ch. 21	Ch. 22, 29

Notes:

(a) Low illuminances (less than 30 lux) are given in lux; values greater than 30 lux are given in letter categories.

(1) Intersections and conflict zones may require higher illuminances.

(2) Lighting must not interfere with visibility for pedestrians, motorists, or boaters.

(3) Hazards such as stairs or areas adjacent to bodies of water should be clearly identified and lighted for safety.

(4) Illuminances for parking areas listed in Chapter 22, Roadway Lighting.

(5) Illuminances for Roadway listed in Chapter 22, Roadway Lighting.

(6) Illuminances for Rest Areas listed in Chapter 22, Roadway Lighting.

(7) Illuminances for Tunnels should provide luminances listed in Chapter 22, Roadway Lighting.

(8) Illuminances for Walkways and Bikeways listed in Chapter 22, Roadway Lighting.

APPLICATIONS GUIDE

Applications Guide

This chapter presents photos of 12 typical interior or exterior lighting scenarios, brief descriptions of the situation and objectives, and guidelines and issues concerning appropriate lighting for the applications.

    Example 1. Daylighted Classroom
    Example 2. Open Office (direct/indirect luminaires with task lighting)
    Example 3. Open Office (workspace-specific "intelligent" pendant luminaires)
    Example 4. Open Office (recessed parabolic luminaires)
    Example 5. Industrial Assembly Space/Manufacturing
    Example 6. Industrial Warehouse
    Example 7. Bank Lobby and Teller Line.
    Example 8. Retail Big Box store
    Example 9. Retail Clothing Store.
    Example 10. Supermarket
    Example 11. Mall Chain Store
    Example 12. Banking Office Entrance and Parking Lot
    Example 13. Courtroom
    Example 14. Audience Chamber for Theatre

Figure 20-1. *(Design credits courtesy of Weller and Michal Architects, Inc. Design-Lights Consortium, Classroom KnowHow. George Leisey Photographer.)*

#1 Daylighted Classroom

Context and Objectives

Lighting for classrooms should support the education experience by providing a comfortable, attractive environment with a focus on the educator.

Issues and Guidelines

- √ *Daylight:* Classrooms are ideally suited for daylighting because the rooms are relatively narrow and are most heavily used during daylight hours. In addition to providing windows for view, consider clerestories, monitors and skylights to introduce daylight from many directions. Provide overhangs, light-shelves, awnings, shades and blinds to control direct solar penetration and glare.

- √ *Task Visibility (Source/Task/Eye Geometry and Reflected Glare):* Rooms should be laid out so that the windows are to the side of the student for most activities. Computer screens should be facing away from windows to avoid reflected glare. Choose a lighting system that minimizes reflections on computer screens and locate luminaires close to the black/white board to minimize reflections.

- √ *Color Appearance:* Glass should be selected so that it retains the color spectrum of daylight to the greatest extent possible. The color of electric lighting should be selected primarily for use in evening classes. CRI should be above 80 to ensure natural appearance of skin tones, educational materials, clothing and finishes.

- √ *Visual Comfort (Discomfort Glare and Overhead Glare):* Avoid direct solar penetration. Provide window and skylight brightness control. Since the occupants are in the heads-up position for much of the time, focusing on the teacher or boards, special consideration should be made for overhead glare. Avoid excessively bright lamps in luminaires where the lamps are visible.

- √ *Room Surface Brightness:* Illuminate ceiling and wall surfaces to make the workspace appear brighter and more cheerful. Use light-colored materials, 70% reflectance on walls, and 80% reflectance on ceilings and window walls, to encourage inter-reflections and reduce contrast between windows, skylights and their adjacent surfaces. Use saturated or medium tone colors only below 0.75 meters (30 inches) above the finished floor, or as limited accents.

- √ *Light Distribution on Surfaces:* Illuminate walls and ceilings in a relatively uniform manner. Provide more light—about three times the ambient light level—on the primary black/white board to increase visibility and make it a focal point for students.

Figure 20-2. *(Photo courtesy of the Lighting Research Center.)*

#2 Open Office (Direct/Indirect Luminaires and Task Lighting)

Context and Objectives

This office space is connected to a high-tech manufacturing facility, and the architect wanted to express the industrial character of the building. Worker satisfaction is a high priority for the owner. The work performed in the space is computer-intensive.

Issues and Guidelines

√ *Room Surface Brightness:* Illuminate ceiling and wall surfaces to make the workspace appear brighter and more cheerful. Use light-colored materials, 70% reflectance on walls, and 80% reflectance on ceilings and window walls to encourage inter-reflections and reduce contrast between windows, skylights and their adjacent surfaces. Use saturated or medium tone colors only below 0.76 meters (30 inches) above finished floor, or as limited accents.

√ *Light Distribution on Surfaces:* To avoid distracting reflections on computer screens, specify a system with some uplight where the maximum ceiling luminance is no more than four times the minimum ceiling luminance. Indirect luminaires with some downward glow are recommended to reduce the contrast of an opaque luminaire against an illuminated ceiling.

√ *Daylight and View:* Windows and skylights can provide significant benefits to this space but can be powerful sources of glare for computer screens. Design skylights in workspaces so that no direct sunlight strikes the workplane. Provide overhangs, lightshelves, awnings, shades, blinds, or curtains to block workers' direct view of the sun.

√ *Task Visibility (Source/Task/Eye Geometry and Reflected Glare):* Position computer screens so that windows are not reflected in the screen; also select a lighting system that will minimize reflections on computer screens. (In this installation, a low-ambient illuminance direct/indirect fluorescent system worked well.) For lighting mounted under overhead cabinets, specify optical systems designed to minimize veiling reflections on shiny paper tasks. If the ambient light levels are relatively low, provide the option of a compact fluorescent or LED movable desktop task luminaire for areas of the desk that are not lighted by an undercabinet luminaire.

√ *Visual Comfort (Discomfort Glare and Overhead Glare):* Block users' direct view of bright lenses or bare lamps in task or overhead luminaires. Louvers and baffles are an effective way to block direct view of bare lamps from these angles. Avoid using direct luminaires with extremely bright exposed lamps (e.g., T5 fluorescent lamps) even when louvers are used, because they can be a source of overhead glare. Provide window brightness controls.

√ *Color Appearance:* Keep CRI of light sources above 80 to ensure pleasant appearance of skin tones, clothing and finishes.

Figure 20-3. *Office Lighting Research Study Space, Albany, New York. (Photo courtesy of Light Right Consortium.)*

#3 Open Office (Advanced Controls and Personal Dimming)

Context and Objectives

Advances in technology allow the use of digital addressable ballasts in a layered controls approach to optimize energy savings and flexibility. This approach is characterized by direct/indirect luminaires specifically located at each workstation. Lamps for uplight and downlight are circuited separately to allow personal dimming control of the downlight component as well as the use of occupancy sensors. The uplight component is not under the control of the user, which ensures uniform brightness on the ceiling plane and enables grouped control of these lamps for peak load shedding, lumen maintenance and perimeter daylight dimming. The connected load is reduced due to the proximity of the luminaires to the workstations. This approach provides improved comfort and satisfaction for occupants and can significantly reduce energy use.[92, 93]

Issues and Guidelines

- √ *Room Surface Brightness:* Illuminate ceiling and wall surfaces to make this space appear brighter and more cheerful. With this type of design, the direct-indirect luminaires are positioned with the workstations and may end up some distance from the corridor walls. Where this happens, specify a wallwash system to ensure walls are not too dark. Undercabinet task luminaires serve the dual function of illuminating the vertical partitions and reducing shadows, and can usually do the job with reduced-output ballasts. Use light-colored materials for room surfaces, 70% reflectance on walls, and 80% reflectance on ceilings and window walls to encourage interreflections and reduce contrast between windows, skylights and their adjacent surfaces. Use saturated or medium tone colors only below 0.76 meters (30 inches) above finished floor, or as limited accents.

- √ *Light Distribution on Surfaces:* To avoid distracting reflections on computer screens, most pendant luminaires should be suspended at least 38 centimeters (15 inches)—preferably 45 centimeters (18 inches) or greater—from the ceiling to optimize ceiling luminance ratios. If workstations are unusually large, the suspension may need to be increased or luminaires added to avoid excessively dark spaces between the luminaires. Workstation-specific luminaires are designed to fit within the typical cubicle workspace. If there are no vertical partitions between workstations, then there is a risk that dimming from different workstations can impact neighbors. Verify that workstations are far enough apart and/or partitioned to avoid user conflict over dimming preferences.

- √ *Daylight:* Integrate the electric lighting with daylight to allow the uplight portion of the perimeter fixtures to be dimmed. As with all office spaces, it is important to be careful about windows and skylights creating glare for computer screens. Provide overhangs, light-shelves, awnings, shades, blinds, or curtains to block users' direct view of the sun. Design skylights in workspaces so that no direct sunlight strikes the task plane.

- √ *Task Visibility (Source/Task/Eye Geometry and Reflected Glare):* Position computer screens so that windows are not reflected in them; select a lighting system that will minimize reflections on them. For lighting mounted under overhead cabinets, specify optical systems designed to minimize veiling reflections on shiny paper tasks. Where storage bins occur, under-cabinet task lighting is necessary to reduce shadows on the horizontal desk surface and to provide light for the vertical partition. Other areas of the desktop do not need additional task lighting because the overhead lighting is controllable by the occupant.

- √ *Visual Comfort (Discomfort and Overhead Glare):* Be careful about daylight as a source of glare; provide window brightness controls. In order to reduce overhead glare, choose luminaires with baffles or louvers on the downward side that block view of bright, bare lamps.

- √ *Color Appearance:* Keep CRI of light sources above 80 to ensure pleasant appearance of skin tones, clothing and finishes.

#4 Open Office (Direct Lighting Using Recessed Parabolic Luminaires)

Context and Objectives

The goal of lighting this office space was to control reflected glare on computer screens without creating a cave-like space for users. Parabolic-louvered luminaires can create scallops at the upper surfaces of walls and darker areas between luminaires, unless they are placed close to the walls or supplemented with a dedicated wallwash system.

Issues and Guidelines

- √ *Task Visibility (Source/Task/Eye Geometry and Reflected Glare):* If direct-only lighting is used, ensure that no tall workstation panels or overhead cabinets will block light from overhead luminaires, particularly if the luminaires are widely spaced. Select a lighting system and locations that will minimize reflections on computer screens. (In this installation, fluorescent luminaires with deep-cell semi-specular parabolic louvers worked well.) An even better solution may be to convince the client to purchase improved computer screens such as high-luminance, flat LCD screens with low specularity or anti-reflection screen coatings.

- √ *Visual Comfort (Discomfort Glare and Overhead Glare):* Select luminaires for general lighting that limit luminous intensity at normal viewing angles. (The luminaire should limit light emitted between 60° and 90°, but avoid highly polished metal louvers because they can contribute to a cave-like space appearance.) Louvers and baffles are an effective way to block the direct view of bare lamps from these angles. Avoid using luminaires with extremely bright lamps (e.g., T5HO fluorescent lamps), even when louvers are used, because they can be a source of overhead glare.

- √ *Room Surface Brightness:* Illuminate ceiling and wall surfaces to make the space appear brighter and more cheerful. If the ceiling cannot be lighted, light the walls to avoid a gloomy, cave-like effect. (This installation's designer selected semi-specular parabolic louver luminaires and spaced them within 0.6-0.9 meters, or 2-3 feet, of the walls.) Use light-colored materials, 70% reflectance on walls, and 80% reflectance on ceilings and window walls to encourage inter-reflections and reduce contrast between windows, skylights and their adjacent surfaces. Use saturated or medium tone colors only below 0.76 meters (30 inches) above finished floor, or as limited accents.

- √ *Color Appearance:* Keep CRI of light sources above 80 to ensure pleasant appearance of skin tones, clothing and finishes.

Figure 20-4. *(Photo courtesy of Cooper Lighting.)*

#5 Industrial Assembly Area/Manufacturing

Context and Objectives

Manufacturing and other industrial spaces can experience extremes in temperature, physical size and cleanliness conditions. Adverse conditions, affecting both users and lighting equipment, are common. The issues below apply to industrial spaces in general, although it should be kept in mind that specific types of industry will have additional requirements and restraints.

Issues and Guidelines

√ *Task Visibility (Source/Task/Eye Geometry and Reflected Glare):* A manufacturing employee may be working with specular (shiny) finish materials that can reflect the image of bright luminaires overhead, making it difficult to detect small details. A properly located luminaire can be an aid in inspecting a specular surface for flaws. If the reflections are a nuisance or hindrance, changing the relative positions of the luminaires and the visual task can help. It may also be helpful to use luminaires with a more diffuse light distribution so that any created reflections are less objectionable.

√ *Visual Comfort (Discomfort Glare and Overhead Glare):* In a space with a low ceiling (less than 5 meters, or 16 feet), consider luminaires that shield HID and high-output (HO) fluorescent lamps from direct view. This may be accomplished with reflectors or prismatic optical systems. Larger optical surfaces spread the lamp image over a larger area, thereby

Figure 20-5. *(Photo courtesy of Holophane.)*

decreasing luminance and apparent brightness. If upward viewing is part of this work, HID luminaires should be completely enclosed with an optical assembly that mitigates bare lamp brightness. Exposed fluorescent T8 and T8HO lamps generally do not present glare problems, but exposed T5HO lamps may. Ceiling luminance can mitigate this problem by reducing the contrast between the luminaire and its background. Using luminaires with some uplight and a high ceiling reflectance can be a good solution.

√ *Shadows:* Manufacturing spaces should be lighted in a uniform manner, as shadows will usually hinder work and productivity. Tall equipment makes it difficult to avoid shadows if the luminaires are spaced too far apart. Luminaires may need to be spaced much closer together than their published Spacing Criteria suggest if tall equipment is present. (Caution: Most luminaire layouts are created with lighting software that assumes an empty space.)

√ *Room Surface Brightness:* Light-colored room surfaces (walls, ceiling and even the floor) serve two major roles. First, they reflect most of the light that strikes them, supplementing the direct light in the space and filling in shadows. Less energy is used since less light is absorbed by the room surfaces. Second, lighter-colored surfaces decrease the contrast between the luminaires and their background, increasing visual comfort for workers.

√ *Horizontal and Vertical Illuminance:* The task plane in manufacturing usually occurs in three dimensions, not just on a horizontal tabletop, with workers standing at a lathe, press, cutter, paint sprayer or other vertical equipment. The manufactured objects are often three-dimensional as well; for workers to see them well, the light must be distributed to all planes in the space, vertical, horizontal, and everywhere in between. This is usually accomplished with luminaires that have a broad light distribution.

√ *Modeling of Faces or Objects:* Often, the visual tasks involve seeing objects that are three-dimensional in nature. The shape, contours and surface details must be seen during both manufacturing and quality inspection. Illumination that is too directional (such as straight down) can cause deep shadows that make it difficult to see. On the other hand, if the illumination is too uniform on all planes, three-dimensional features will not stand out well. A horizontal-to-vertical illuminance ratio of 2:1 usually enables good modeling of objects.

√ *Intrinsic Material Characteristics:* Often, workers need to see surface characteristics as a part of the manufacturing process. These include how shiny or matte a finish is; its color, both hue and consistency; how smooth or grainy it is; and other characteristics. The position of a task luminaire can enhance or hinder the visibility of these characteristics, depending on the type of task light and its position relative to the visual task and the worker's line of sight. The IESNA's Recommended Practice for Lighting Industrial Facilities, RP-7-01,[94] has specific information on available task lights, which types of industrial visual tasks they are appropriate for, and how to best locate them.

√ *Color Appearance and Color Contrast:* For some processes, color appearance can be important or critical. Select a light source with a high CRI rating (keeping in mind the limitations of CRI as a metric). Depending on the application, a CRI above 80 might be sufficient. In some cases, a CRI above 90 will be required. Daylight, some fluorescent, and some metal halide lamps are possible sources. While halogen and other incandescent sources have CRI ratings nearing 100, they are shorter lived, less efficacious, and weaker on the blue-violet end of the spectrum than other types.

√ *Flicker and Strobe*: Magnetically-ballasted fluorescent and HID lamps operated at 60Hz AC frequency actually turn on and off 120 times per second. A small percentage of people are sensitive to this flicker, especially in their peripheral vision. Machinery moving back and forth or rotating at a multiple of the flicker rate can appear to be motionless or turned off, posing a strobe effect hazard for workers. Eliminate flicker and strobe by using high-frequency electronic ballasts (20 kHz and higher). Alternatively, operate magnetically-ballasted luminaires on sequential phases of a three-phase electrical supply. When one luminaire is off, its neighbors are on and will fill in for it.

#6 Industrial Warehouse

Context and Objectives

Industrial warehouse spaces have large geometries and varied conditions. Similar to industrial manufacturing spaces, adverse conditions for both users and lighting equipment are common.

Warehouse applications present several different kinds of visual tasks, including reading labels on aisles, bins and products; accessing the material either by hand or using a forklift; counting small parts out of a bin; and reading paperwork or keypad entries.

The issues below apply to industrial warehouse spaces in general, although it should be kept in mind that specific types of warehouses will pose additional requirements and restraints.

Figure 20-6. *(Photo courtesy of the Lighting Research Center.)*

Issues and Guidelines

√ *Color Appearance and Color Contrast:* Sometimes labels on shelves or on stored goods are color-coded, or sometimes the objects themselves have colors that help to identify and distinguish them. The light source in these situations should have a CRI of at least 60. If discerning subtle color differences is important, the CRI should be at least 80.

√ *Room Surface Brightness:* In a warehouse, the "room surfaces" include the faces of the warehouse shelves and the product that is stored on them. If those surfaces are dark in color, they will absorb light, and the lighting design must allow for these lower reflectances. If possible, the ceiling should be painted white to a) minimize the contrast of the luminaires against the ceiling, thereby increasing visual comfort, and b) to reflect light down into the aisles, increasing vertical illuminance and reducing shadows. Similarly, a light-colored floor will reflect light up onto the faces of the lower shelves.

√ *Task Visibility (Source/Task/Eye Geometry and Reflected Glare):* If packages are shrink-wrapped or covered in some other shiny material, reflections can make reading labels or other markings difficult. This is less likely with diffuse sources, such as enclosed prismatic fluorescent or HID luminaires. Their reflections are softer and may be less obscuring than a reflection from a bright lamp in an open luminaire.

√ *Visual Comfort (Discomfort Glare and Overhead Glare):* The warehouse worker may need to be able to read the labels or otherwise identify product on the top shelves. His or her upward gaze should not be directed into an open-reflector HID luminaire. The glare from the exposed lamp will make vision difficult or impossible. If upward viewing is part of this work, HID luminaires should be completely enclosed, with an optical assembly that reduces the brightness of the bare lamp or flashed reflector. If fluorescent luminaires are used, exposed T8 and T8HO lamps are generally not a problem, but exposed T5HO lamps can be.

√ *Light Distribution on Surfaces:* Contrary to popular belief, the visual task in a warehouse is not seeing the floor. It is reading the labels on the faces of the shelves and on the products themselves. These tasks are typically located in a vertical plane and vary in height from centimeters (inches) above the floor to the top of the highest product on the highest shelf, and from one end of the aisle to the other. (For these reasons, standard horizontal illuminance lighting calculation methods do not apply to warehouse lighting.) Vertical illuminances should be computed on a vertical plane that corresponds to the inside edge of the aisle shelving. It is important that the lighting on this visual task be as uniform as possible—typically not easy in a long, narrow space. However, anywhere that a label or other means of identifying the products can occur, there must be sufficient lighting to read that label. Not just on the middle shelves or beneath the luminaires. Once the vertical surfaces have enough light, the floor will too—the reverse, however, is not the case.

√ *Daylight:* The use of daylight in a warehouse space can reduce the amount of electric light (and energy) that is needed. Daylight sensors should be located so as to measure the actual amount of light on the task plane in individual aisles and control the electric luminaires accordingly. When the warehouse aisles are laid out, aisles should be aligned with the skylights for best daylight performance.

#7 Bank Lobby with Teller Line

Context and Objectives

This application consists of bank teller counters and a lobby for customers. Tellers review documents, count money and review computer screens for account information. A security camera records the transaction and the customer.

Issues and Guidelines

- √ *Task Visibility (Source/Task/Eye Geometry and Reflected Glare):* The teller deals with many paper tasks that may use highly reflective ink. Locate downlights so that veiling reflections do not impair visibility for both either the teller or the customer. Suggest to the bank that they can circumvent lighting problems on computer screens by purchasing high-luminance flat-panel LCD displays or high-luminance screens with anti-reflection coatings. The screen orientation should be adjustable so that tellers can tilt or tweak the screen to avoid annoying screen reflections.

- √ *Color Appearance:* Use a light source with a CRI above 80, both for pleasant rendering of skin tones, and potentially for enhancing the teller's ability to spot counterfeit currency.

- √ *Light Distribution on the Task Plane (Uniformity):* Keep the ratio of maximum-to-minimum illuminance on the teller counter within a 2:1 ratio or else it may be difficult for the teller to see details on papers in the darker areas.

- √ *Facial Modeling:* Avoid using harsh downlighting on the faces of the teller or customer. Some diffuse light from suspended luminaires or reflected light from ceiling and wall surfaces will help soften facial shadows. This has the potential of improving speech intelligibility because it is easier to understand speech if one can see facial features moving.

- √ *Vertical Illuminance:* The installed security camera requires a minimum of 100 lx (10 fc) of illuminance on the face of the customer.

- √ *Visual Comfort (Discomfort glare and Overhead Glare):* Avoid the use of bare-lamp downlights directly above the heads of teller and customer.

- √ *Daylight:* Use overhangs, blinds, shades or drapes on windows to control daytime glare.

- √ *Appearance of Space and Luminaires:* The lighting should enhance the architectural features of the space and correspond to the layout of the teller station.

Figure 20-7. *(Photo courtesy of Cooper Lighting.)*

#8 Big Box Discount Store

Context and Objectives

Big box discount stores feature very high ceilings and high aisles. The space is typically windowless and non-partitioned, with a low level of architectural detail. Discounted merchandise is typically displayed in large quantities without much differentiation between products. The lighting strategy is similar throughout the store because the product displays are fairly uniform.

Issues and Guidelines

- √ *Color Appearance:* Sources should have greater than 80 CRI. The color temperature is typically on the warm side (3000K-3500K), but color temperature can vary by preference.

- √ *Appearance of Space and Luminaires:* Surface reflectances are important in these types of stores. Due to the ceiling height, the ceiling should be a light reflectance so the space does not feel cave-like. Luminaires are typically utilitarian in their appearance, to be in concert with the affordable merchandise.

- √ *Daylight:* Most big box spaces are windowless, but toplighting and skylighting are becoming more common. Where toplighting is used, photosensors should be integrated with the electric lighting system to dim or step-dim the general lighting system.

- √ *Visual Comfort (Discomfort Glare):* Luminaires are typically at high mounting heights, which can reduce the discomfort glare issues, as long as the light sources are not overly bright. Consider the brightness of the luminaires as they will appear against the ceilings and walls.

- √ *Light Distribution on Task Plane (Uniformity):* In this type of space, the task plane is primarily the vertical surface of the racks where merchandise is displayed. Light distribution for the merchandise racks should be as uniform as possible. Additionally, light distribution should be fairly uniform on the horizontal plane for feature displays, checkout areas and in the aisles.

- √ *Vertical Illuminance:* Ensure that light levels are adequate towards the bottom of the racks.

Figure 20-8.

#9 Retail Clothing Space

Context and Objectives

Retail environments range from big box to high end. The ultimate goal of every owner, however, is to maximize profitability from sales of merchandise in the store. The owner should carefully consider the lighting scheme and its effect on sales revenue and energy and maintenance expenses.

Issues and Guidelines

√ *Appearance of Space and Luminaires:* In this store, the cove uplighting helps show off the architectural shell as well as providing ambient lighting that provides vertical illuminance on faces, bodies and merchandise.

√ *Color Appearance:* Consider fluorescent lamps in cove lighting and halogen lamps in track lighting that deliver light with a CRI greater than 80.

√ *Visual Comfort (Discomfort Glare):* Track-mounted directional luminaires accent the merchandise. The tracks are spaced no more than one mounting height apart, giving the staff the chance to aim luminaires so that they are directed between 0° and 30° from nadir (i.e., more downward than outward), reducing possibilities for discomfort glare for shoppers looking into the lamps. Additionally, check the cutoff angles of light emitted from the recessed fixtures with parabolic louvers to determine if this is appropriate for the space (keep in mind that people are often looking at or above the horizontal in a retail space). In other words, can people see the lamps in a recessed fixture when they are trying to look at merchandise?

√ *Light Distribution on Surfaces:* All horizontal and vertical surfaces need not be uniformly lighted. Variation in illuminances will establish a visual hierarchy of merchandise in the store and is central to the design strategy. Illuminance on featured merchandise should typically be 3-5 times greater than the general ambient illuminance for transition and non-sale areas.

√ *Modeling of Faces or Objects:* It is equally critical that people look good to others, such as when shopping with friends or family, or to themselves, such as when trying on clothing in dressing rooms. The intensity and angles of light will determine whether the light is flattering to the merchandise and the people in the store. Light modeling is critical in fitting rooms where people hope to look 10 pounds thinner and 10 years younger; flattering lighting can help sell merchandise. Avoid intense, directional lighting in these areas.

Figure 20-9. *(photographer: John Sutton Photography)*

CHAPTER 20: APPLICATIONS GUIDE

Figure 20-10. *(Photo courtesy of Holophane.)*

#10 Supermarket

Context and Objectives

In supermarkets, ceilings are medium to high in height, and aisles are medium height. They are typically windowless non-partitioned spaces with a low level of architectural detail. The lighting strategy is similar throughout most of the store, but there are certain products that require higher color rendering and accent lighting, such as product, meats and floral displays. Due to the prevalence of fresh food products in supermarkets, it is important that accent lighting be appropriately designed to ensure that heat gain on the food does not pose a problem.

Issues and Guidelines

- √ *Color Appearance:* Lamp CRI ratings should be greater than 80.

- √ *Appearance of Space and Luminaires:* Surface reflectances are important in these types of stores. Due to the ceiling height, the ceiling should have a light reflectance so the space does not feel cave-like.

- √ *Visual Comfort (Discomfort Glare):* Luminaires are typically at high mounting heights, which can reduce discomfort glare issues, as long at the light sources are not overly bright. Consider the brightness of the luminaires as they will appear against the ceilings and walls. Accent lighting should be well shielded so it is not glaring for customers. For horizontal displays such as vegetable bins, the aiming angles need to be 30º (as measured from vertical) or less to avoid glare. For vertical shelving displays, the aiming angles should be 45º or less from vertical.

- √ *Light Distribution on Task Plane (Uniformity):* In this type of space, a primary task plane is the vertical surface of the racks where merchandise is displayed. Light distribution for the merchandise racks should be as uniform as possible. Vegetables, fruit and meat displays are typically found in bins, so the task plane is sloped or horizontal and below eye level. Accent lighting is appropriate for these bins.

- √ *Task Visibility (Source/Task/Eye Geometry and Reflected Glare):* Reflections from the overhead luminaires are often seen in the glass of the refrigerator and freezer case doors. This increases the need for internal casework lighting so the merchandise is visible. Although often relegated to the casework manufacturer, the designer should attempt to control lamp type, location and color rendering as part of the design process.

- √ *Shadows:* Shadows are a common problem inside casework with deep solid shelving. Task lighting, internal to the casework, can be helpful. If the shelves are fixed, then the casework lighting should be mounted under the lip of each shelf and angles should be shallow enough to reduce shadows.

- √ *Vertical Illuminance:* It is especially important to ensure that light levels are adequate towards the bottom of the racks.

#11 Mall Chain Store

Context and Objectives

Mall chain stores typically have ceiling heights ranging from 3 to 4 m (10 to 13 ft). This example shows a small- to medium-size chain store where accent lighting is often close to half or more of the lighting design. The ratio of accent to ambient lighting varies depending on the store brand and type of product. In a typical mall, anchor stores (department stores) are typically quite large and the lighting strategy is similar to supermarkets, where the ambient lighting prevails with selected areas of accent lighting. By contrast, smaller mall chain stores depend heavily on accent lighting to attract consumer attention and to differentiate products. Window display lighting is also a high priority in malls, where the mall concourse is the area of greatest competition.

Issues and Guidelines

- √ *Appearance of Space and Luminaires:* These types of stores have more visual variety than big box or department stores, and there is a higher contrast between accent/display and ambient lighting. Often, there is a considerable amount of fixed perimeter display, but the floor displays are often variable in relation to seasonal changes. Architectural finishes are often more refined and detailed, often expressing the brand identity.

- √ *Color Appearance:* Light sources should have greater than 80 CRI at a minimum, and 90 CRI is recommended when possible. Color temperature is typically warm, around 3000-3500K. When using different types of lamps, it is important for color temperatures and color rendering to be very similar.

- √ *Visual Comfort (Discomfort Glare and Overhead Glare):* Accent lighting is often achieved with adjustable track or ceiling-mounted luminaires. It is important to make sure that aiming angles are steep enough that they do not shine directly into the eyes of customers. Use louvers and baffles to provide appropriate shielding.

Figure 20-11. *South Coast Plaza Gap Store.*

√ *Modeling of Faces or Objects:* Modeling of items for sale is extremely important in these types of stores. Consumer decisions are made based on how well they can see the merchandise.

√ *Light Distribution on Surfaces:* Most surfaces are opportunities to showcase merchandise, including wall displays, racks, shelves and gondolas. It is not necessary to have a uniform distribution of light throughout the store—in fact, much of the design strategy in this type of store is related to varying luminance ratios. Illuminance on featured merchandise should typically be 3-5 times greater than the general ambient illuminance for transition and non-sale areas. In very high-end stores, there can sometimes be very little ambient light, and contrast ratios can be 10:1 or higher.

√ *Task Visibility (Source/Task/Eye Geometry and Reflected Glare):* For horizontal retail displays, aiming angles need to be 30° or less (as measured from straight downward) to avoid glare. For vertical merchandise or shelving displays, aiming angles should be no more than 45°. See Visual Comfort, above.

√ *Sparkle (or Desirable Reflected Highlights):* The type of light source can be a major factor in revealing the attributes of the merchandise. Point sources do a better job of revealing sparkle, texture and reflected highlights compared to diffuse sources.

√ *Horizontal and Vertical Illuminance:* The ambient illuminance needs to be sufficiently low to allow the accent lighting to visually emphasize the merchandise while meeting energy codes.

Figure 20-12. *(lighting design: Naomi Miller; photography: © Randall Perry Photography, LLC)*

#12 Banking Office Entrance and Parking Lot

Context and Objectives

This building houses a corporate banking office, teller lobby and ATM. Employees or customers are present at all times of day, with vehicles dropping off individuals at the entrance or parking. User safety is a concern at almost all hours after dark.

Issues and Guidelines

- √ *Peripheral Detection:* Illuminate building façades, alleyways, areas around trees and shrubs, fences, walls and anywhere criminals can find concealment and opportunity. Consider lamps rich in short (blue) wavelengths but avoid light spill into bedroom windows if any section of the walkway is located near a residential area.

- √ *Color Appearance:* Color is important for identifying and describing people, clothing, cars, etc. For parking and pathway lighting, use light sources with CRI ratings greater than 50. Color appearance is also important for landscaping appearance.

- √ *Task Visibility (Disability Glare):* Sometimes a parking lot or pathway light is so bright that a pedestrian can only see the silhouette of the approaching person, making it impossible to identify the person by their facial features or clothing. In this project, full-cutoff pole-mounted metal halide luminaires mounted at 10m (32 ft.) and spaced 31m (100 ft.) apart were used to provide even lighting on the roadway and parking lot surface. Compact fluorescent wall sconces mounted on the columns direct light downward along the curb line. Both of these emit little light above 80°, so glare does not reduce visibility.

- √ *Modeling of Faces:* Vertical illuminance is critical for facial modeling, but luminaires emitting high-angle light are also glaring. Uniformity of light on the horizontal surface (a 4:1 average: minimum ratio or less) is related to good vertical illuminance. In areas where facial modeling is critical for safety, it may be necessary to space poles closer together than the luminaire's Spacing Ratios would suggest. (Usually this allows a reduction in lamp wattage.) Avoid using low-height bollards when facial recognition is important. These emit little light upward onto faces, and any light directed upward will also appear glaring.

√ *Horizontal Illuminance:* Follow IESNA guidelines for parking lot and walkway illuminance. You may be able to reduce levels slightly if the outdoor area is in a rural area, where ambient light levels are lower; conversely, it may be appropriate to raise them in high-ambient-illuminance urban areas.

√ *Vertical Illuminance:* Ask: Can pedestrians see at a significant distance? Vertical illuminance enables pedestrians to identify surroundings and should be a design priority. It is important to provide vertical illuminance without creating a glare source.

√ *Light Distribution on Task Plane (Uniformity):* Follow spacing criteria to keep the average-to-minimum ratio of illuminance on the parking and pathway surface within a 4:1 ratio to improve the pedestrian's ability to see details along the road or walkway surface. Also consider the uniformity of light at a height of 1.5m (5 ft.) above the path so as to promote good facial modeling.

√ *Shadows:* Locate luminaires so that they will produce a minimum of mysterious shadows from trees, shrubs, walls and other features. This may entail mounting the luminaires below the canopy of the trees or trimming the lower branches of the trees.

#13 Courtroom

Context and Objectives

Courtroom environments should evoke respect for authority and tradition. The activities in a courtroom have a significance and gravity that should be reflected in the design decisions. Lighting decisions must be thoughtfully made in this type of environment because the consequences of improper lighting are more significant than in most other environments. Visual tasks include reviewing evidence, watching testimony and, increasingly, audio-visual presentations. The lighting system must be responsive to the variety of needs while supporting the architectural choices.

Issues and Guidelines

- √ *Modeling of Faces or Objects:* It is critically important to be able to clearly perceive both objects and faces in courtrooms. To accomplish this objective, use a layered approach for the lighting design and provide both direct and indirect lighting. Direct lighting helps viewers to perceive the subtleties of the evidence and the testimony, while the indirect lighting provides bounced light and interreflections to minimize harsh shadows and contrast.

- √ *Appearance of Space and Luminaires:* Courtrooms are often high-ceilinged spaces, with impressive and sometimes majestic architectural solutions. The lighting should support the architectural design intent, which often means integrating the lighting through the use of architectural coves, hidden wallwashing and other means to create visual simplicity. When luminaires are visible, as in the example shown, they are usually traditional and attractive in their appearance, often with a luminous element and an indirect distribution.

- √ *Color Appearance:* Lamp colors for the different lighting systems should be approximately the same to minimize potential for confusion. For example, if one of the lighting systems is dimmed or increased for the purposes of evidence viewing, the color appearance of the objects and people should not change to a noticeable degree. All light sources should have a CRI rating higher than 80.

- √ *Daylight:* Use overhangs, blinds, shades or drapes on windows to control daytime glare.

Figure 20-13. *(Photo courtesy of Matt Franks, Arup Lighting.)*

- √ *Visual Comfort (Discomfort Glare and Overhead Glare):* It is essential to ensure that the judge, jury and lawyers do not experience glare from windows or downlight luminaires. In high-ceiling courtrooms, the downlight luminaires are typically at high mounting heights, which can reduce the glare as long at the light sources are not overly bright. Consider the luminaire brightness as it will appear against the ceilings and walls.

- √ *Task Visibility (Source/Task/Eye Geometry and Reflected Glare):* Video presentations must be clearly visible to the jurors, so pay attention to the relationship between the light sources, the video display screen and the viewing angles to ensure that glare does not obscure the screen.

- √ *System Control and Flexibility:* Many different tasks occur in courtrooms, requiring that the lighting system be flexible and controllable. It is essential to be able to reduce light on projection screens during presentations, for example; as a result, careful consideration should be given to the various control and dimming options. While halogen infrared PAR lamps are inexpensive to dim, they are also energy-intensive. In a high-ceiling space, the wattage needed for PAR lamp downlights is typically far too high to meet energy codes. Consider dimming fluorescent lamps instead. Metal halide lamps can be used in a layered approach with dimming fluorescent lamps—but *only* if the newest technology is used. Pulse-start and ceramic metal halide offer significant performance improvements over traditional probe-start metal halide, and electronic ballasts are available for some metal halide lamps, optimizing system performance and efficiency. The controls for the lighting system must be easy to understand and operate as well as accessible to the judge and other courtroom staff as required.

- √ *Special Considerations:* In addition to video presentations, it is increasingly common for videotaping and filming to occur in courtrooms. In this situation, vertical illuminances must be significantly increased without creating discomfort or glare for the occupants. Additionally, depending on jurisdiction, special guidelines may apply related to controls, emergency lighting and security lighting.

- √ *Horizontal and Vertical Illuminance:* Follow IESNA recommendations for horizontal illuminance. If videotaping or filming is planned, follow IESNA recommendations for vertical illuminance for video recording.

- √ *Room Surface Brightness:* One of the most common problems in traditional courtrooms is the excessive use of dark wood as a finish for walls. While this is a common, traditional finish in courtrooms, its use compromises other important aspects of the design. Light-colored room surfaces should be used rather than wood because they reflect most of the light that strikes them, supplementing the direct light in the space and filling in otherwise problematic shadows. In addition, less energy is used because less light is absorbed by room surfaces.

Figure 20-14. *Robert and Margrit Mondavi Center. (lighting design: Auerbach + Glasow; photographer: Robert Canfield, San Rafael, CA.)*

#14 Audience Chamber for Theatre

Context and Objectives

Theatres are often used as auditoriums or classrooms as well as for enjoying performances, The lighting should be designed to accommodate different functions. Theatrical luminaires for stage lighting and architectural luminaires for the audience chamber should be coordinated so that they do not interfere with each other.

Issues and Guidelines

- √ *Room Surface Brightness:* Colors of surfaces are usually medium or dark in value. If room surfaces are white or very light in color, light from the stage can make these surfaces bright in appearance, distracting from the central focus of the stage.

- √ *Appearance of Space and Luminaires:* In a historic theatre with decorative ornament covering room surfaces, architectural decorative luminaires can be an important part of the décor in the space. Theatrical luminaires and non-decorative architectural luminaires most likely need to be hidden in architectural slots or coordinated with the décor to minimize their appearance. In high-tech or modern-looking theatre, the theatrical luminaires themselves might become an important decorative feature.

- √ *Points of Interest:* Highlights on decorative elements or decorative luminaires themselves may create sparkle or points of interest. Notice the small decorative sconces attached to the face of the balcony in the photograph.

- √ *Color Appearance:* In some theatres, daylight is available for its classroom function but is completely blocked out for the performance function. In other theatres, there are no windows, so layers of electric light with different lamps are provided to accommodate different functions. Materials or paint colors should be selected under each of these various lamps.

√ *Task Plane Illuminance:* Before the performance starts, the audience should be able to find a seat and be able to read a program. During the performance, the audience chamber should be dark enough to make the stage the center of focus. At the same time, the aisles should be lighted sufficiently for people to find their way safely to and from seats without distracting others. Higher illuminances need to be provided for functions such as cleaning the space or taking final exams.

√ *Task Visibility (Source/Task/Eye Geometry and Shadows):* Pronounced sharp shadows from the body or hand can be very distracting, particularly when the space is used as a classroom with a test-taking function. Diffused light should be used to minimize sharp shadows on exam pages.

√ *System Control and Flexibility:* Because the space is used for different functions, it usually has separately dimmed or switched layers of light. Layers may include downlights for audience task light before and after performances, decorative chandeliers or sconces, step lights and aisle lights for egress paths, and an energy-efficient light source to provide higher light levels for classroom use or cleaning, etc.

√ *Maintenance:* Since the ceiling of the audience chamber is usually high, luminaires in the ceilings are often accessed from above the ceiling, or when they are suspended, they are lowered close to the floor. It is very important to plan ahead to provide access to the luminaires for changing lamps, ballasts or transformers.

√ *Energy Use:* Finding an energy-efficient lighting solution that works well for multiple functions in a high-ceiling space is challenging. It is important to evaluate the strengths and weaknesses of different lamp types and design approaches to determine the best way to compose layers of light for each project.

√ *Special Considerations (Luminaire Noise):* For a concert hall or other space where sound is critical, luminaires that are used for the performance function must be quiet. See Chapter 12 for more information about luminaire noise.

APPENDIX I.
GLOSSARY OF TERMS

Adaptation: The process by which the retina becomes accustomed to more or less light than it was exposed to during an immediately preceding period. It results in a change in the sensitivity to light. Note that adaptation is also used to refer to the final state of the process, such as reaching a condition of adaptation to a specific luminance level.

Ambient Lighting: Lighting throughout an area that produces general illumination that may or may not be uniform.

Ballast: A device used with an electric discharge lamp to obtain the necessary circuit conditions (voltage, current, and waveform) for starting and operating.

Brightness: The attribute by which an area of finite size (such as a wall or the luminous portion of a luminaire) is perceived to emit, transmit, or reflect a greater or lesser amount of light. No judgment is made as to whether the light comes from a reflecting, transmitting, or self-luminous object.

Correlated Color Temperature (of a lamp) (CCT): The absolute temperature of a blackbody radiator having a chromaticity (or color) equal to that of the lamp.

Contrast: The relationship between the luminance of an area and the luminance of its immediate background. The difference between those luminances divided by the luminance of the background. See Luminance Contrast.

Color Rendering Index (of a lamp) (CRI): A measure of the degree of color shift objects undergo when illuminated by the lamp as compared to those same objects when illuminated by a reference source of comparable color temperature.

Exitance (luminous exitance): $M = d\phi/dA$. The area density of luminous flux leaving a surface at a point. Note that this is the total luminous flux emitted, reflected, and transmitted from the surface and is independent of direction.

General Lighting: Lighting designed to provide a substantially uniform level of illuminance throughout an area, exclusive of any provision for special local requirements.

Illuminance: $E = d\phi/dA$. The area density of luminous flux incident at a point on a surface, measured in lumens per square meter or lux (lx), or lumens per square foot or footcandles (fc).

Lamp: A generic term for a source created to produce optical radiation. By extension, the term is also used to denote sources that radiate in regions of the spectrum adjacent to the visible. Note that through popular usage, a portable light fixture consisting of a lamp with shade, reflector, enclosing

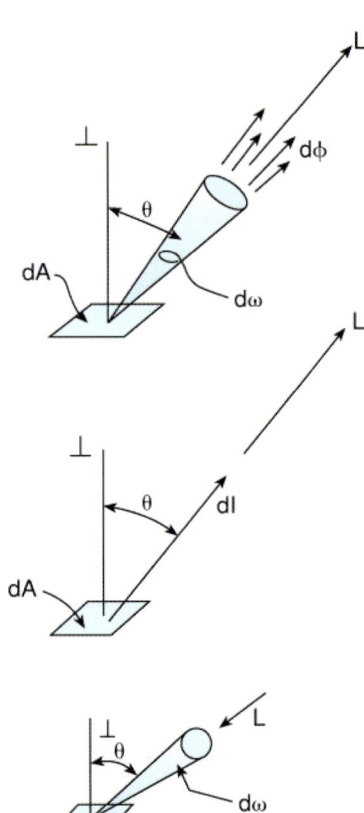

Figure A-1. *(156) Three ways to express luminance.*

globe, housing, or other accessories is also sometimes called a "lamp," but is more accurately called a "luminaire."

Luminaire (lighting fixture): A complete lighting unit consisting of a lamp or lamps and ballast(s) (when applicable) together with the parts designed to distribute the light, to position and protect the lamps, and to connect the lamps to the power supply.

Luminance (formerly photometric brightness): $L = d^2\phi/(d\omega\, dA \cos\theta)$ (in a direction and at a point of a real or imaginary surface). The quotient of the luminous flux at an element of the surface surrounding the point, and propagated in directions defined by an elementary cone containing the given direction, by the product of the solid angle of the cone and the area of the orthogonal projection of the element of the surface on a plane perpendicular to the given direction. The luminous flux can be leaving, passing through, and/or arriving at the surface. By introducing the concept of *luminous intensity*, luminance can be expressed as $L = dI/(dA \cos\theta)$. Here, luminance at a point on a surface in a direction is interpreted as the quotient of luminous intensity in the given direction, produced by an element of the surface surrounding the point, by the area of the orthogonal projection of the element of surface on a plane, perpendicular to the given direction. Luminance can be measured at a receiving surface by using $L = dE/(dA\cos\theta)$. This value can be less than the luminance of the emitting surface due to the attenuation of the transmitting media.

Note that in common usage, the term "brightness" usually refers to the strength of sensation that results from viewing surfaces or spaces from which light comes to the eye. This sensation is determined in part by the definitely measurable luminance defined above and in part by conditions of observation such as the state of adaptation of the eye. In much of the literature, brightness, when used alone, refers to both luminance and sensation. The context usually indicates which meaning is intended. Previous usage not withstanding, neither the term brightness nor the term photometric brightness should be used to denote the concept of luminance.

Luminance Contrast: The relationship between the luminances of an object and its immediate background. It is equal to $(L_1 - L_2)/L_1$ or $(L_2 - L_1)/L_1 = |\Delta L/L_1|$, where L_1 and L_2 are the luminances of the background and object, respectively. The form of the equation must be specified. The ratio $\Delta L/L_1$ is known as Weber's fraction. See note under *luminance*. Because of the relationship among luminance, illuminance, and reflectance, contrast often is expressed in terms of reflectance when only reflecting surfaces are involved. Thus, contrast is equal to $(\rho_1 - \rho_2)/\rho_1$ or $(\rho_2 - \rho_1)/\rho_1$ where ρ_1 and ρ_2 are the reflectances of the background and object, respectively. This method of computing contrast holds only for perfectly diffusing surfaces; for other surfaces it is only an approximation unless the angles of incidence and view are taken into consideration. See Reflectance.

Reflectance (of a surface or medium): $\rho = \varphi_r/\varphi_i$. The ratio of the reflected flux to the incident flux. Reflectance is a function of:

> 1. Geometry
>   a. of the incident flux
>   b. of collection for the reflected flux
> 2. Spectral distribution
>   a. characteristic of the incident flux
>   b. weighting function for the collected flux
> 3. Polarization
>   a. of the incident flux
>   b. component defined for the collected flux.

Note that unless the state of polarization for the incident flux and the polarized component of the reflected flux are stated, it should be considered that the incident flux is unpolarized and that the total reflected flux (including all polarization) is evaluated. Spectral reflectance depends on only the beam geometry and the character of the reflecting surface (and on polarization). Luminous reflectance also is a function of the spectral distribution of the incident flux. If no qualifying geometric adjective is used, the reflectance for hemispherical collection is meant. Certain of the reflectance terms are theoretically imperfect and are recognized only as practical concepts to be used when applicable. Physical measurements of the incident and reflected flux are always biconical in nature. Directional reflectances cannot exist, since one component would be finite while the other was infinitesimal; here the reflectance distribution function is required. However, the concepts of directional and hemispherical reflectance have practical application in instrumentation, measurements, and calculations when including the effect of the nearly zero or nearly 2π conical angle would increase complexity without appreciably affecting the immediate results. In each case of conical incidence or collection, the solid angle need not be a right cone but can be of any cross section, including a rectangle, a ring, or a combination of two or more solid angles. For many geometrically specified reflectance properties, it is assumed that the radiance (luminance) is isotropic over the specified solid angle of incidence. Otherwise, the property is a function of the directional distribution of the radiance (luminance) as well as the beam geometry and the character of the reflecting surface.

Spectral Power Distribution Chart: A chart published by lamp manufacturers or laboratories showing the relative proportion of colors radiated. For example, a cool white lamp shows a strong blue at 450 nanometers where the warm white shows a strong red at 650 nanometers.

Task Plane: Specific surface or area on which visual attention is placed.

Veiling Reflection: Specular reflections that are superimposed upon diffuse reflections from an object that partially or totally obscure the details to be seen by reducing the contrast. This sometimes is called *reflected glare*. Another kind of veiling reflection occurs when one looks through a plate of glass. A reflected image of a bright element or surface can be seen superimposed on what is viewed through the glass plate.

Visible Spectrum: All the colors of the rainbow. This is the portion of the electromagnetic spectrum lying between 380 and 760 nanometers.

Volumetric Brightness (room surface brightness): The perceived response to the luminances of room surfaces and surfaces within a room. Volumetric brightness can improve the appearance of cheerfulness. The light-colored surfaces reflect diffuse light back into the room, which can add vertical illuminance to faces and objects that makes them easier to see and reduces harsh shadows.

APPENDIX II. ENDNOTES

[1] IESNA, *IESNA Lighting Handbook,* IESNA, New York. 2000.

[2] Rea, M. S. 1986. Toward a model of visual performance: Foundations and data. *J. Illum. Eng. Soc.* 15(2):41-57.

[3] Boyce, P. R., and M. S. Rea. 1987. Plateau and escarpment: The shape of visual performance. *Proceedings: 21st session. Commission Internationale de l'Éclairage*, Paris: Bureau Central de la CIE.

[4] Baron, R. A., M. S. Rea, and S. G. Daniels. 1992. Effects of indoor lighting (illuminance and spectral distribution) on the performance of cognitive tasks and interpersonal behaviors: The potential mediating role of positive affect. *Motiv. Emot.* 16(1):1-33.

[5] Kaplan, S. 1987. Aesthetics, affect, and cognition: environmental preference from an evolutionary perspective. *Environment and Behavior* 19:3-32.

[6] Kaplan, S., and R. Kaplan. 1982. *Cognition and the environment: Functioning in an uncertain world*. Ann Arbor : Ulrich's.

[7] Loe, D. L., K. P. Mansfield, and E. Rowlands. 1994. Appearance of lit environment and its relevance in lighting design: Experimental study. *Light. Res. Tech.* 26(3):119-133.

[8] Loe, D. L., and E. Rowlands. 1996. The art and science of lighting: A strategy for lighting design. *Light. Res. Tech.* 28(4):153-164.

[9] Rowlands, E., D. L. Loe, R. M. McIntosh, and K. P. Mansfield. 1985. Lighting adequacy and quality in office interiors by consideration of subjective assessment and physical measurement. *CIE Journal* 4(1):23-37.

[10] Wilkins, A. J., I. Nimmo-Smith, A. I. Slater, and L. Bedocs. 1989. Fluorescent lighting, headaches and eyestrain. *Light. Res. Tech.* 21(1):11-18.

[11] Wilkins, A. J., I Nimmo-Smith, A. I. Slater, and L. Bedocs, 1989. Fluorescent lighting, headaches and eyestrain. *Lighting Research and Technology* 21 (1):11-18.

[12] Stevens, R. G., B.W. Wilson, and L. E. Anderson, eds. 1997. *The melatonin hypothesis: Breast cancer and use of electric power*. Columbus OH: Battelle Press.

[13] Silverstone, B., Lang, M.A., Rosenthal, B.P. and Faye, E.E., *The Lighthouse Handbook on Vision Impairment and Vision Rehabilitation*. Oxford University Press, New York. 2000.

[14] Joan E. Roberts, Light interactions with the human eye as a function of age. Lighting Research Office International Symposium Proceedings, Orlando, FL, Lighting and Human Health, 2002.

[15] P.R. Boyce, *Illumination,* in The Handbook of Human Factors and Ergonomics, ed: G. Salvendy, John Wiley and Sons, New York, 1997.

[16] Newsham, G.R.; Arsenault, C.D.; Veitch, J.A.; Tosco, A.M.; Duval, C.L. "Task lighting effects on office worker satisfaction and performance, and energy efficiency," Leukos, 1, (4), 2005 http://irc.nrc-cnrc.gc.ca/fulltext/nrcc48152/

[17] IESNA RP-1 2004 *Office Lighting.*

[18] *Advanced Lighting Guidelines*. New Buildings Institute, 2002.

[19] Boyce, P.R., Hunter, C.M., and Inclan, C., Overhead Glare and Visual Discomfort, Journal of Illuminating Engineering Society, of North America, P.73, Winter 2003.

[20] Lighting For Quality—Overhead Glare, LD+A, February, 2003.

[21] Sheedy, J. E., and Bailey, I.L., Symptoms and Reading Performance with Peripheral Glare Sources, Work with Display Units 94, Eds. A Grieco, G. Molteni, B. Piccoli, and E. Occhipinti, Elsevier Science, Amsterdam, 1995

[22] Ngai, P., and Boyce, P.R., The Effect of Overhead Glare on Visual Discomfort, Journal of Illuminating Engineering Society, P.29, summer 2000.

[23] Commission Internationale de l'Eclairage [CIE]. *Discomfort glare in interior lighting* (Publication No. 117-1995). Vienna, Austria CIE, 1995.

[24] IESNA *Recommended Practice: Lighting for Exterior Environments,* RP-33-99, New York: IESNA, 1999.

[25] IESNA *Recommended Practice: Lighting for Parking Facilities,* RP-20-98, New York: IESNA, 1998.

[26] IESNA *Recommended Practice: Roadway Lighting,* ANSI/IESNA RP-8-00 (Reaffirmed 2005), New York: IESNA, 2000.

[27] Carter, C.B., and Boyce P.R., "Lighting standards for videoconferencing in distance learning environments" *Journal of the Illuminating Engineering Society,* Vol. 32, No. 2, 2003. pp. 37-51.

[28] Stanley McCandless. *A Method of Lighting the Stage*, 4th edition. New York: Theatre Arts Books. 1932 - 1958.

[29] Guo, X., and Houser, K.W., "A review of colour rendering indices and their application to commercial light sources" *Lighting Research and Technology,* Vol. 36, No.3, 2004 pp. 183-200.

APPENDIX II: ENDNOTES

[30] Lighting Research Center, *Lighting Answers: Light Sources and Color* Rensselaer Polytechnic Institute, Troy, New York, 2004.

[31] Quellman, E., and Boyce, P.R., "The Light Source Color Preferences of People of Different Skin Tones", *Journal of the Illuminating Engineering Society*, Vol. 31, No. 1, 2002. pp. 109-118.

[32] Waide, P., Tanishima, S. et al, *Light's Labour's Lost: Policies for Energy-Efficient Lighting*, International Energy Agency, 2006.

[33] Mills, E., "Global lighting energy savings potential," *Light & Engineering*. 2002. Vol 10, No 4, pp 5-10.

[34] *Advanced Lighting Guidelines*, 2003 Edition. New Buildings Institute, White Salmon, WA. 2003.

[35] Consortium for Energy Efficiency. CEE High-Performance Commercial Lighting Systems Initiative, Qualifying Products List, High Performance 4' T8 Lamps. http://www.cee1.org/com/com-lt/com-lt-prod.pdf, October 2005.

[36] "Fluorescent and other Mercury-Containing Lamps and the Environment: Mercury Use, Environmental Benefit, Disposal Requirements", NEMA, March 2005 http://www.nema.org/gov/ehs/committees/lamps/upload/Lamp%20Brochure%20Final%203%2005.DOC as of 4/26/05

[37] *Advanced Lighting Guidelines*. New Buildings Institute, 2002.

[38] International Energy Agency (IEA) Solar Heating and Cooling Programme, Energy Conservation in Buildings & Community Systems. (2000). Daylight in Buildings: A Source Book on Daylighting Systems and Components (Report of IEA SHC Task 21 / ECBCS Annex 29, July 2000). Berkeley, CA: Lawrence Berkeley National Laboratory. Available at: http://gaia.lbl.gov/iea21/ieapuba.htm

[39] Veitch, J.A.; Newsham, G.R. 2000. "Preferred luminous conditions in open-plan offices: research and practice recommendations," Lighting Research and Technology, 32, (4), pp. 199-212.

[40] Newsham, G.R.; Veitch, J.A.; Arsenault, C.D.; Duval, C.L. "Effect of dimming control on office worker satisfaction and performance," Proceedings of IESNA Annual Conference (Tampa, Florida), pp. 19-41, 2004. URL: http://irc.nrc-cnrc.gc.ca/fulltext/nrcc47069/)

[41] Newsham, G.R.; Veitch, J.A.; Arsenault, C.D.; Duval, C.L. "Effect of dimming control on office worker satisfaction and performance," Proceedings of IESNA Annual Conference (Tampa, Florida), pp. 19-41, 2004. URL: http://irc.nrc-cnrc.gc.ca/fulltext/nrcc47069/)

[42] Boyce PR, Veitch JA, Newsham GR, Myer M, Hunter C. *Lighting quality and office work: A field simulation study*, A report for the Light Right Consortium, December 2003. (Available at www.lrc.rpi.edu and http://irc.nrc-cnrc.gc.ca/fulltext/b3214.1/)

[43] Newsham, G.R.; Veitch, J.A.; Arsenault, C.D.; Duval, C.L. "Effect of dimming control on office worker satisfaction and performance," Proceedings of IESNA Annual Conference (Tampa, Florida), pp. 19-41, 2004. URL: http://irc.nrc-cnrc.gc.ca/fulltext/nrcc47069/

[44] Figueiro, M. 2004. Occupancy sensors: are there reliable estimates of energy savings? LD+A, January, p 8-9.

[45] Jennings, J.D.; Rubinstein, F.M.; DiBartolomeo, D.; Blanc, S.L. 2000. "Comparison of control options in private offices ...", Journal of the Illuminating Engineering Society, Summer, pp 39-60. URL: http://eetd.lbl.gov/btp/papers/43096REV.pdf

[46] National Electrical Manufacturers Association (NEMA), *Guidelines on the Application of Dimming to High Intensity Discharge Lamps*, LSD 14-2002, 2002.

[47] Website link http://www.darksky.org/links/enviro.html (accessed 2005-02-21)

[48] IESNA *Recommended Practice: Roadway Lighting*, ANSI/IESNA RP-8-00(Reaffirmed 2005), New York: IESNA, 2000.

[49] IESNA *Recommended Practice: Lighting for Exterior Environments*, RP-33-99, New York: IESNA, 1999.

[50] IESNA *Recommended Practice: Lighting for Parking Facilities*, RP-20-98, New York: IESNA, 1998.

[51] IESNA *Technical Memorandum on Light Trespass: Research, Results, and Recommendations*, TM-11-00, New York: IESNA 2000.

[52] IESNA, *Addressing Obtrusive Light (Urban Sky Glow and Light Trespass) in Conjunction with Roadway Lighting*, TM-10-00, New York: IESNA 2000.

[53] Akashi, Y. and Rea, M, 2002. Peripheral detection while driving under a mesopic light level, Journal of the Illuminating Engineering Society, New York.

[54] Terman, M., Lewy, A.J., Dijk, D-J., Boulos, Z., Eastman, C.I., and Campbell, S.S., "Light treatment for sleep disorders: Consensus report. IV. Sleep phase and duration disturbances", *Journal of Biological Rhythms*, Vol. 10, No. 2, 1995. pp. 135-147.

[55] Commission Internationale de L'Eclairage [CIE], Ocular Lighting Effects on Human Physiology and Behaviour, Vienna, Austria, 2004, p. 6-7.

56 He, Y., Rea, M.S., Bierman, A. and Bullough, J., "Evaluating light source efficacy under mesopic conditions using reaction times" *Journal of the Illuminating Engineering Society*, Vol. 26, No. 1, 1997. pp. 125-138.

57 Newsham, G.R.; Sander, D.M. "The Effect of office design on workstation lighting: a simulation study," Journal of the Illuminating Engineering Society, Summer 2003, pp. 52-73. URL: http://irc.nrc-cnrc.gc.ca/fulltext/nrcc45357/

58 LaGiusa, F., and Perney, L. R., "Further studies on the effects of brightness variations on attention span in a learning environment" *Journal of the Illuminating Engineering Society*, Vol. 3, 1974. pp. 249-252.

59 Myer, M., Boyce, P.R., and Akashi, Y., Sparkle: Perception and Production *Proceedings of the Illuminating Engineering Society of North America, Annual Conference*, Tampa, July, 2004. pp. 305-333.

60 Myer, M., Boyce, P.R., and Akashi, Y., Sparkle: Perception and Production *Proceedings of the Illuminating Engineering Society of North America, Annual Conference*, Tampa, July, 2004. pp. 305-333.

61 Akashi, Y., Y. Tanabe, I. Akashi, and K. Mukai. 2000. Effect of sparkling luminous elements on the overall brightness of impression: A pilot study. *International Journal of Lighting Research and Technology* 32, (1) 19-26.

62 Akashi, Y. 1999. Sparkle elements—a bright idea. *Architectural Lighting* 14: 38-41.

63 Barr, V and Welantaz, B. 1978. Lighting the Marketplace: Sparkle, fantasy, and merchandising. Lighting Design and Application (LD&A) 8 (10) 29-33.

64 Slater, A.I., and Boyce, P.R., "Illuminance uniformity - where's the limit" *Lighting Research and Technology*, Vol 22, No. 4, 1990. pp. 165-174.

65 IESNA, *Lighting Handbook,* 9th Edition, IESNA, New York. 2000.

66 Boyce PR, Veitch JA, Newsham GR, Myer M, Hunter C. *Lighting quality and office work: A field simulation study*, A report for the Light Right Consortium, December 2003. (Available at www.lrc.rpi.edu and http://irc.nrc-cnrc.gc.ca/fulltext/b3214.1/)

67 Newsham, G.; Arsenault, C.; Veitch, J.; Tosco, A.M.; Duval, C. "Task Lighting Effects on Office Worker Satisfaction and Performance, and Energy Efficiency." *Leukos*, v. 1, no. 4, April 2005, pp. 7-26.

[68] Newsham, G.; Veitch, J.; Arsenault, C.; Duval, C. 2004. "Lighting for VDT Workstations 2: Effect of Control and Lighting Design on Task Performance, and Chosen Photometric Conditions" Research Report IRC-RR-166, Institute for Research in Construction, National Research Council Canada, Ottawa http://irc.nrc-cnrc.gc.ca/fulltext/rr166/].

[69] Newsham, G.R.; Richardson, C.; Blanchet, C.; Veitch, J.A. "Lighting Quality Research using Rendered Images of Offices", Lighting Research & Technology, v. 37, no.2, 2005, pp. 93-115. url: http://irc.nrc-cnrc.gc.ca/fulltext/nrcc47072.

[70] Miller, N., Boyce P., and MacKay, H. *An approach to a measurement of lighting quality*, Proceedings of the Illuminating Engineering Society of North America Annual Conference, 1995

[71] Boyce, P.R., and Slater, A.I., "The application of CRF to office lighting design" Lighting Research & Technology, Vol. 13, No. 2, 1981. pp. 65-79.

[72] Farley, K.M.J.; Veitch, J.A. "A Room With A View: A Review of the Effects of Windows on Work and Well-Being", Research Report, Institute for Research in Construction, National Research Council Canada, IRC-RR-36, 33 pages, 2001. URL: http://irc.nrc-cnrc.gc.ca/fulltext/rr136/.

[73] Heschong, L.; Wright, R.L.; Okura, S., "Daylighting impacts on human performance in school", JIES 2002 31(2) Summer 101-114.

[74] Boyce, P. R. The Benefits of Daylight Through Windows", available at www.lrc.rpi.edu).

[75] Keighley, E. C. (1973). Visual requirements and reduced fenestration in offices: a study of multiple apertures and window area. *Building Science, 8*, 321-331.

[76] Keighley, E. C. (1973). Visual requirements and reduced fenestration in offices a study of window shape. Building Science, 8, 311-320.

[77] Webb, Ann. Ultraviolet and Vitamin D: Essential Exposure. CIE Light and Health Symposium. 2004. Vienna.

[78] Science News, vol. 166, pages 232-233. October 9, 2004.

[79] International Energy Agency (IEA) Solar Heating and Cooling Programme, Energy Conservation in Buildings & Community Systems. (2000). Daylight in Buildings: A Source Book on Daylighting Systems and Components (Report of IEA SHC Task 21 / ECBCS Annex 29, July 2000). Berkeley, CA: Lawrence Berkeley National Laboratory. Available at: http://gaia.lbl.gov/iea21/ieapuba.htm

APPENDIX II: ENDNOTES

80   ASHRAE – American Society of Heating, Refrigerating and Air-Conditioning Engineers. 2007. Energy Standard for Buildings Except Low-Rise Residential Buildings. ASHRAE/IESNA/ANSI 90.1-2007. ASHRAE. Atlanta, GA.

81   IESNA *Recommended Practice: Lighting for Exterior Environments*, RP-33-99, New York: IESNA, 1999.

82   Boyce, P.R. *Human Factors in Lighting (2nd Edition)* Taylor and Francis. London. 2003.

83   Pease, K., "A review of street lighting evaluations: Crime reduction effects", in *Crime Prevention Studies*, eds. K. Painter and N. Tilley, Criminal Justice Press, Monsey, NY. 1999.

84   Boyce, P.R., Eklund, N.H., Hamilton, B.J., and Bruno, L.D. "Perceptions of safety at night in different lighting conditions" *Lighting Research and Technology*, Vol. 32, No. 2, 2000. pp. 79-92.

85   IESNA Recommended Practice: Lighting for Exterior Environments, RP-33-99, New York: IESNA, 1999.

86   IESNA *Recommended Practice: Roadway Lighting*, ANSI/IESNA RP-8-00 (Reaffirmed 2005), New York: IESNA, 2000.

87   Ouellette, M. J., B. W. Tansley, and I. Pasini. 1993. The dilemma of emergency lighting: Theory vs reality. J. Illum. Eng. Soc. 22(1):113-121.

88   Rea, M.S., Clark, F.R.S., and Ouellette, M.J., *Photometric and psycho-physical measurements of exit signs through smoke*, DBR paper 1291, National Research Council Canada, Ottawa. 1985.

89   Lighting Research Center, *Specifier Reports: Exit Signs, Troy*, New York, LRC, 1994.

90   IESNA RP-30-96, Recommended Practice on Museum and Art Gallery Lighting

91   IESNA, *IESNA Lighting Handbook,* IESNA, New York. 2000

92   Boyce PR, Veitch JA, Newsham GR, Myer M, Hunter C. *Lighting quality and office work: A field simulation study*, A report for the Light Right Consortium, December 2003. (Available at www.lrc.rpi.edu and http://irc.nrc-cnrc.gc.ca/fulltext/b3214.1/)

93   Jones, C., Gordon, K. Efficient Lighting Design and Office Worker Productivity. 2004ACEEE Summer Study in Buildings Proceedings. American Council for an Energy Efficient Economy, Washington, DC.

94   IESNA Recommended Practice for Lighting Industrial Facilities, ANSI/IESNA RP-7-01. IESNA. New York, NY.

INDEX

A
Accent lighting 87
ADA. *See* Americans with Disabilities Act
Adaptation 159
Aesthetic judgment, in lighting design 4–5
    dimensions of appraisal in 5
    surface brightness in 5
    visual interest in 5
Aging, disability glare and 15
Albinism 5
Ambient lighting 159
Americans with Disabilities Act (ADA), in U.S. 111
ANSE. *See* Associón de Normalización y Certificación
Aragon, Louis 25
Architecture, lighting design in 7–8, 78–84
    downlighting in 81
    finish choices in 81
    furniture finish/sizes and 82
    integration of 80
    light patterns in 84, 88
    luminaire location in 81
    luminaire selection in 79–81
    mounting height in 81
    project-related issues with 125
    spacing in 79–80, 82–83
Area sources, of lighting 36
ASHRAE/IESNA Standards, for energy codes 109–110
Associón de Normalización y Certificación (ANSE) 110
Atmosphere 4
    from light patterns 85
ATMs. *See* Automated Teller Machines
Auditoriums, lighting design for 157–158
Automated Teller Machines (ATMs), lighting for 30–31, 153

B
Ballast 159
    output 56
Banks, lighting design for 146
    at ATMs 30–31, 153
    for entrances/parking lots 153–154
Big box stores, lighting design for 147
Brain, in human visual system 11
Brightness 159
    from light patterns 86
    luminance v. 12, 14
    room surface 94–99
    volumetric 95, 162
Browning, Elizabeth Barrett 39
Building codes 110, 112, 114
    ADA and 111
    ANCE and 110
    in Canada 110, 112
    CSA 110
    in Mexico 110, 114
    MNECB 111–112
    NBC 112
    NFPA and 110
    in U.S. 110

C
Cameras, lighting for 119
Canada 110–112
    building codes in 110, 112
    CSA in 110
    energy codes in 111–112
    MNECB in 111–112
    NBC in 112
Canadian Standards Association (CSA) 110
Cancer, artificial light and 5
Candelas 12–13
    luminance and, as measurement unit for 14, 98
CCT. See Correlated color temperature
Circadian rhythms 102
Classrooms, lighting design for 134–135
Coding. *See* Energy codes
Cognitive dissonance, from light patterns 89
Color contrast 18
Color, light and 39–46
    appearance of 39
    cool sources for 40
    neutral sources for 40
    quality of 39
    rendering 6, 39, 42–46
    temperature 6, 39–41

  visible spectrum of 162
  warm sources for 40
Color rendering 6, 39, 42–46
  CRI for 42–46, 159
  under fluorescent lighting 44
  under halogen luminaires 44
  lighting source influence on 44–45
  purpose of 42
  for skin tones 43
  SPD in 45, 161
Color rendering index (CRI) 42–46, 159
  for commercial spaces 46
  for lamps 43
  for lighting design 42–46
  measurement scale for 43
  object test colors in 43
  for workspaces 45–46
Color temperature 6, 39–41
  CCT and 41, 159
  kelvins and 39, 41
  in lighting design 6
Columbus, Christopher 3
Commercial spaces, lighting design for
  banks 146
  big box stores 147
  controls in 64–68
  CRI in 46
  emergency egress in 117–118
  entrances/parking lots 153–154
  light distribution in 86–89
  light pollution from 72
  mall chain stores 151–152
  offices 136–140
  retail clothing stores 148
  sparkle in 91
  supermarkets 149–150
Computers, glare from 21
Cones, in human visual system 11
Constable, John 33
Contrast 14–18, 159
  color 18
  disability glare and 15
  luminance 15
  shadows and 16
  source/task/eye geometry influence on 14
Controls, in lighting design 63–68
  automated 67
  commercial use of 64–68
  dimming/switching strategies for 64–66, 68
  for energy 63
  for energy use 63
  for flexibility 64–65
  for light pollution 72
  localized/centralized 67
  mood through 65
  for occupant satisfaction 65
  in offices 138–139
  threshold-based events as 63
  time-based event as 63
  zone definition/number in 65–66
Le Corbusier 79
Cornea, in human visual system 11
Correlated color temperature (CCT) 41, 159
Costs, of lighting design 49–50
  budgeting for 49
  energy-efficient lighting and 50
  initial 49
  life-cycle 49
  maintained 49
  maximizing value of 49–50
Courtrooms, lighting design for 155–156
Cradle-to-cradle design 60
CRI. *See* Color rendering index
CSA. *See* Canadian Standards Association

D

Daylight 101–108
  diffuse 101
  direct, control of 104
  electric light and 107–108
  glare in 103–104
  harvesting of 101
  luminance ratios with 103
  window glare in 104
Daylight harvesting 101
Daylighting 101–103, 105–107
  advantages of 101–103
  design strategies for 105–107
  disadvantages of 103
  energy conservation with 108
  human circadian rhythms and 102
  room shape and 105–107

	with toplighting	106
	window configurations and	107

Diffuse daylight ... 101
Diffuse reflectance ... 97
    See also Matte finishes, lighting and
Diffused lighting ... 35, 37
Dimmers ... 7
    in lighting design controls ... 64–66
    NEMA guidelines for ... 68
    in office lighting design ... 138–139
Direct daylight, control of ... 104
Directional lighting, for visual tasks ... 17
Disability glare, aging and ... 15
Discomfort glare ... 25–26
    sparkle in ... 25
Distribution of light. *See* Light distribution
Downlighting ... 35
    in architectural lighting design ... 81
    uniformity in light distribution and ... 92

E

Emergency egress, lighting design for ... 117–118
    in World Trade Center ... 119
Emerson, Ralph Waldo ... 69
Emotional response, from lighting design ... 4–5
Energy codes ... 109–114
    See also Building codes
    ASHRAE/IESNA Standards for ... 109–110
    in Canada ... 111–112
    EPAct ... 111
    IEC ... 109
    LPD in ... 109
    in Mexico ... 112–113
    NEC ... 110
    in U.S. ... 109–110
Energy Policy Act (EPAct), in U.S. ... 111
Energy Saving Fund (FIDE) ... 113
Energy use, in lighting design ... 55–58
    ballast output and ... 56
    controls for ... 63
    daylight harvesting and ... 101
    energy-efficiency and ... 55
    environmentally-sensitive ... 7
    global operating costs of ... 55–56
    light source efficacy and ... 55–56
    lighting controls and ... 57
    photometric distribution of lamps and ... 56–57
    room reflectances and ... 57
Energy-efficient lighting ... 50, 55–57
    daylighting as ... 108
    lighting controls and ... 57
    in maintenance of lighting designs ... 53
    room reflectances and ... 57
    in workspaces ... 58
Environmental issues, in lighting design ... 59–61
    cradle-to-cradle design and ... 60
    with lamps ... 60–61
    LEED and ... 60
    local customization for ... 60
    product life-cycles as ... 60
    rating systems and ... 60
    recycling and ... 61
    sustainability as ... 59–60
EPAct. *See* Energy Policy Act
Ewing, Sam ... 51
Exitance ... 98–99, 159
Eye, in human visual system ... 11–14
    components of ... 11
cones in ... 11
    illuminance range for ... 13
    luminance range for ... 14
    retina ... 11
    rods in ... 11

F

Facial modeling ... 34–36
    under diffused lighting ... 35
    under downlighting ... 35
    shadow use in ... 36
FIDE. *See* Energy Saving Fund
Field of view ... 26
    horizontal/vertical ... 26
    overhead glare and ... 26
Flexibility controls, in lighting design ... 64–65
Flicker ... 30–31
    definition of ... 30
    in lamps ... 31
    minimization tactics for ... 31
    strobe effects and ... 30
    visual comfort and ... 25, 30–31
Fluorescent lighting, color rendering under ... 44

Footcandles 13
        See also Lux
Foster, Alan Dean 115

G
Glare 25–31
        from computer screens 21
        in daylight 103–104
        disability 15
        discomfort 27–28
        elimination tactics for 27–30
        in lighting for safety 117
        in night lighting 70–74
        overhead 27–28
        sources of 27
        from track lighting 29
        Unified Glare Ratings for 27
        window 104
Green design 59
        See also Environmental issues, in lighting design; Sustainability

H
Hakuta, Ken 49
Halogen luminaires 23
        color rendering under 44
        in modeling of objects 34
Henry, O. 63
Hugo, Victor 1
Human needs, from lighting design 4, 6
Human visual system 11
        See also Vision
        color contrast in 18
        cornea in 11
        eye in 11

I
IEC. *See* International Energy Code
Illuminance 9, 13, 20, 159
Incandescent luminaires 23
Industrial spaces, lighting design for 94, 141–143
        warehouses 144–145
Initial costs, in lighting design 49
International Energy Agency 55
International Energy Code (IEC) 109

K
Kahn, Louis 77
Kelvins 39, 41

L
Lamp lumen depreciation (LLD) 51
Lamps 22, 159–160
        CRI ratings for 43
        environmental issues with, in lighting design 60–61
        flicker in 30
        mercury in 61
        replacement of, in lighting design maintenance 52–53
        SPD for 45, 161
        TCLP test for 61
LDD. *See* Luminaire dirt depreciation
Leadership in Energy & Environmental Design (LEED) 60
LEED. *See* Leadership in Energy & Environmental Design
Life-cycle costs, in lighting design 49
Light bulb 22
        See also Lamps
Light distribution 85–100
        with accent lighting 87
        in commercial spaces 86–89
        light patterns in 84–90
        points of interest and 87
        reflectance in 97–98
        room surface brightness and 94–99
        sparkle and 90–91
        on surfaces 85–89
        uniformity in 92–94
        on workplane 92–94
Light measurements 12–13
        candelas 12–13
        footcandles 13
        illuminance in 13
        lumen-hours and 12
        lumens 12
        lux 13
Light patterns 84–90
        architecture and 84, 88
        atmosphere from 85
        brightness perception as result of 86
        cognitive dissonance from 89
        cohesiveness for 89
        in lighting design 89–90
        points of interest from 87
        safety and 89

INDEX

    space perception and 86
Light pollution 69–74
    from commercial structures 72
    controls in lighting design for 72
    factors for 69
    luminaires and 71–74
    minimization tactics for 70–74
Light power density (LPD) 109
Light quality. *See* Quality of light
Light trespass 69–74
    minimization tactics for 70–74
Lighting design. *See also* Architecture, lighting design in; Commercial spaces, lighting design for; Controls, in lighting design; Energy use, in lighting design; Light distribution; Modeling, lighting and; Offices, lighting design for; Visual tasks, lighting for
    with accent 87
    aesthetic judgment in 4–5
    in architecture 7–8, 78–84
    atmosphere from 4
    for auditoriums 157–158
    in banks 146
    benefits of 7–8
    for big box stores 147
    building codes and 110, 112, 114
    cancer development and 5
    for classrooms 134–135
    color in 39–46
    color rendering and 6, 42–46
    color temperature and 6, 41
    for conservation of materials, in museums 119
    control in 63–68
    costs of 49–50
    in courtrooms 155–156
    CRI for 42–46
    with daylight 101–108
    daylighting in 101–103, 105–107
    dimming/switching strategies in 7
    for emergency egress 117–118
    emotional response from 4–5
    energy codes and 109–114
    energy consumption in 7
    energy use and 55–58
    for entrances/parking lots 153–154
    environmental considerations in 59–61
    with fluorescent lighting 44
    goals of 5–6
    guide for 127–132
    human needs served by 4, 6
    for industrial spaces 94, 141–145
    lamps in 22
    through light distribution 85–100
    light patterns in 89–90
    light quality in 7
    luminances in 6
    maintenance of 51–53
    for mall chain stores 151–152
    for manufacturing spaces 141–143
    melatonin production and 5
    modeling and 32–38
    mood and 4
    for night 28–29, 69–76
    for offices 136–140
    project preparation for 123–125
    reflectance as factor in 97
    for retail clothing stores 148
    role of 3
    for safety issues 5, 115–120
    social communication and 5
    for sparkle 91
    SPD in 45, 161
    for supermarkets 149–150
    with track lighting 28
    for uniformity 93
    visual comfort and 4, 25–32
    visual comfort from 4
    for visual tasks 3–4, 9–10
Lighting design projects 123–125
    architecture-related issues for 125
    client goals in 123
    costs assessment for 124
    human needs in 123–124
Lighting fixture 22
        See also Luminaires
Lighting, for cameras 119
Lighting Handbook 93, 127
Linear sources, of lighting 36
LLD. *See* Lamp lumen depreciation
LPD. *See* Light power density
Lumen-hours 12
Lumens 12
Luminaire dirt depreciation (LDD) 51
Luminaires 7, 22, 160

in architectural lighting design 79–84
components of 22
halogen 23
incandescent 23
light pollution and 71–74
maintenance of, through cleaning for 52
narrow focus 23
noise of 119–120
in office lighting design 136–137, 140
photometering process in 22–23
Luminance 6, 12–14, 160
brightness v. 12, 14
candelas as measurement for 14, 98
contrast in 15, 160–161
in daylight, ratios of 103
exitance v., in room surface brightness 98–99
overhead glare and 28–29
for sparkle 90
for visual tasks 12
Luminance contrast 15, 160–161
Luminous flux 12
Lux 13

M

Maintained costs, in lighting design 49
Maintenance, of lighting designs 51–53
correct lamp replacement in 52–53
with energy-efficient lighting 53
LDD and 51
LLD and 51
luminaire cleaning for 52
Mall chain stores, lighting design for 151–152
Manufacturing spaces, lighting design for 141–143
Matte finishes, lighting and 17
Melatonin production 5
Mercury, in lamps 61
Mexico 110, 112–114
ANCE in 110
building codes in 110, 114
energy codes in 112–113
FIDE in 113
MNECB. See Model National Energy Code of Canada for Buildings
Model National Energy Code of Canada for Buildings (MNECB) 111–112
Modeling, lighting and 32–38
area sources for 36
with diffused light 35, 37
of faces 34–36
light sources in 34
linear sources for 36
of objects 33
point sources for 36
with shadow 33, 37–38
of surfaces 33–34
Mood
lighting and 4
through lighting controls 65
Moore, Harriet 37
Motion sensors 73
Movement, task visibility and 19
Municipal spaces, lighting design for courtrooms 155–156
Museums, lighting design for 119

N

Narrow focus luminaires 23
National Building Code (NBC) 112
National Electrical Code (NEC) 110
National Electrical Manufacturers Association (NEMA) 68
National Fire Protection Association (NFPA) 110
NBC. See National Building Code
NEC. See National Electrical Code
NEMA. See National Electrical Manufacturers Association
NFPA. See National Fire Protection Association
Night glare 70–74
Night lighting design 31, 69–76
for ATMs 29
driving detection tasks and 74–75
glare in 70
light pollution and 69
light trespass and 69–70
motion sensors in 73
outdoor 70
overhead glare and 29
peripheral lighting in 73–76
purpose of 69

O

Object modeling, shadow in 33
Off-axis foveal vision 76
Offices, lighting design for 136–140
controls in 138–139
with dimmers 138–139

INDEX

    with direct luminaires..140
    with indirect luminaires 136–137
On-axis foveal vision 76
Outdoor lighting
    for entrances/parking lots 153–154
    for night...70
Overhead glare...25–27
    field of view and..26
    location source for..27
    locations source for 27
    luminance levels and..28–29
    during nighttime..29

P

Perimeter lighting..117
Peripheral lighting..73–76
    enhancement of..75–76
    Off-axis foveal vision and..76
    On-axis foveal vision and...76
Photometering...22–23
    lamp distribution and..56–57
Point sources, of lighting...36
Points of interest..87
    from sparkle 90
Pollution. *See* Light pollution

Q

Quality of light..7

R

Recycling..61
Reflectance...97–98, 161
    diffuse 97
    in lighting design 97
    specular 97
    value of 97
    in workspaces..98
Reflection. *See* Veiling reflection
Rendering. *See* Color rendering
Retail clothing stores, lighting design for......................148
Retina..11
Rods, in eye...11
Room surface brightness..94–99
    disadvantages of..96
    luminance v. exitance in..98–99
    studies for 95
    volumetric brightness and..95
Roosevelt, Theodore...63

S

Safety, lighting design for 5, 115–120
    during emergency egress 117–118
    glare in...117
    illuminance uniformity in 116
    light patterns and...89, 115–120
    with perimeter lighting...117
Schweitzer, Albert..59
Scintillation...90
Shadow 16, 37–38
    in facial modeling..36, 37–38
    lighting sources for..38
    in object modeling...33, 37–38
Shiny surfaces. *See* Specular reflectance
Skin tones, color rendering for 43
Sky glow..69
    See also Light pollution
Skylights..106
Smith, Fran Kellogg 37
Social communication 5
Source/task/eye geometry
    contrast influenced by 14
    vision and..9
    in visual tasks, lighting and 9
Sparkle..25, 90–91
    in commercial spaces 91
    functional role of...90
    lighting design for...91
    luminance levels for..90
    points of interest from..90
    scintillation in..90
SPD. *See* Spectral power distribution
Spectral power distribution (SPD)...............................45, 161
Specular reflectance..97
Steichen, Edward...85
Strobe effects 30
Supermarkets, lighting design for................................149–150
Surface brightness...5
Sustainability..59–60
Switchers...7
    in lighting design controls..64–66, 68

T

Taskplanes 92–94, 162
    See also Workplanes
TCLP. *See* Toxicity Characteristic Leaching Procedure test
Threshold-based events, as lighting control 63
Time, visual tasks and 19
Time-based events, as lighting control 63
Toplighting 106
Toxicity Characteristic Leaching Procedure (TCLP) test 61
Track lighting 28
2003 Blackout 70

U

Unified Glare Ratings 27
Uniformity, in light distribution 92–94
    downlighting and 92
    in industrial applications 94
    lighting design for 93
United Nations World Commission on Environment and Development 59
United States (U.S.) 109–111
    ADA in 111
    building codes in 110
    energy codes in 109–110
    EPAct in 111
U.S. Green Building Council 60

V

van der Rohe, Ludwig Mies 47, 101
Veiling reflection 162
Visibility, for visual tasks 3
    color contrasts in 18
    contrasts in 14–18
    directional lighting and 17
    material considerations for 17
    movement and 19
    planning for 19–21
    shadows and 16
    size as factor in 18
    time as factor for 19
Visible light. *See* Luminous flux
Visible spectrum 162
Vision 9–21
    contrast in 14–18
    illuminance and 13
    luminances in 6, 12–14
    Off-axis foveal 76
    On-axis foveal 76
    source/task/eye geometry in 9
    tasks and 9–10
Visual comfort 4, 25–31
    field of view and 26–27
    flicker and 25, 30–31
    glare and 25–31
Visual tasks, lighting for 3–4, 9–10, 18–21
    color contrasts in 18
    contrasts in 14–18
    definition of 9
    illuminance in 9, 13, 20
    luminance for 12
    matte finishes and 17
    movement and 19
    for performance 3–4
    planning for 19–21
    size as factor in 18
    source/task/eye geometry in 9
    time as factor in 19
    for visibility 3
    on workplanes 14
Volumetric brightness 95, 162

W

Warehouses, lighting design for 143–145
Wharton, Edith 121
Windows
    configurations of, daylighting and 107
    glare through 104
Workplanes 14
    light distribution on 92–94
Workspaces
    CRI for 45–46
    energy-efficient lighting in 58
    reflectance in 98
World Trade Center 119